Strengthening Cooperation over Transboundary Groundwater Resources

Groundwater is humanity's most vital supply of freshwater. Freshwater resources contained in aquifer systems globally are two orders of magnitude greater than those found in all rivers, lakes, and other surface freshwaters combined. Moreover, approximately one half of the world's population today is dependent on groundwater for its basic needs.

While these truisms are widely acknowledged, an additional reality is now recognized – many of these subsurface freshwater reservoirs are actually transboundary formations shared amongst two or more nations. In fact, around 600 transboundary aquifers have been identified worldwide. They underlie the territory of nearly every non-island nation and are found in both humid and arid environments. Surprisingly, though, only a handful of transboundary aquifers worldwide are subject to any type of formal or informal international agreement.

Transboundary groundwater resources are now receiving growing attention among the international community. Questions are being raised about how they should be managed, allocated and protected, what rights aquifer riparians can enjoy, and what responsibilities these aquifer states might owe to neighbouring aquifer riparians. This book considers all these various aspects of the management and regulation of transboundary groundwater resources. It also presents case studies and explores recent efforts to engage the international community on best practices and global standards for governing transboundary aquifers. As communities and nations continue to expand their reliance on these critical sources of freshwater, they will have to address such issues and develop governance regimes for these shared subsurface resources.

The chapters in this book were originally published as a special issue of *Water International.*

Gabriel Eckstein is Professor of Law at Texas A&M University and directs its Energy, Environmental, & Natural Resources Systems Law Program. He serves as President of the International Water Resources Association, and Executive Council Member of the International Association for Water Law.

Alice Aureli is Chief of UNESCO's Groundwater Systems and Water for Human Settlements Section. She is also a member of the Secretariat of UNESCO's intergovernmental International Hydrological Programme where, among other tasks, she is responsible for the International Shared Aquifers Resources Management (ISARM) initiative.

Routledge Special Issues on Water Policy and Governance

Edited by: Cecilia Tortajada *(IJWRD) – Institute of Water Policy, Lee Kuan Yew School of Public Policy, NUS, Singapore*, and **James Nickum** *(WI) – International Water Resources Association, France*

Most of the world's water problems, and their solutions, are directly related to policies and governance, both specific to water and in general. Two of the world's leading journals in this area, the *International Journal of Water Resources Development* and *Water International* (the official journal of the International Water Resources Association), contribute to this special issues series, aimed at disseminating new knowledge on the policy and governance of water resources to a very broad and diverse readership all over the world. The series should be of direct interest to all policy makers, professionals and lay readers concerned with obtaining the latest perspectives on addressing the world's many water issues.

Strengthening Cooperation over Transboundary Groundwater Resources
Edited by Gabriel Eckstein and Alice Aureli

Source-to-Sea Management
Edited by Josh Weinberg, Qinhua Fang, Sarantuyaa Zandaryaa, Greg Leslie and James Nickum

Global Water Resources
Festschrift in Honour of Asit K. Biswas
Edited by Cecilia Tortajada and Eduardo Araral

Putting Water Security to Work
Addressing Global Sustainable Development Challenges
Edited by Chad Staddon and Christopher A. Scott

A River Flows Through It
A Comparative Study of Transboundary Water Disputes and Cooperation in Asia
Edited by Selina Ho

Flood Resilience of Private Properties
Edited by Thomas Hartmann, Willemijn van Doorn-Hoekveld, Helena F.M.W. van Rijswick and Tejo Spit

Virtual Water
Implications for Agriculture and Trade
Edited by Chittaranjan Ray, David McInnes and Matthew R. Sanderson

Legal Perspectives on Bridging Science and Policy
Edited by Mara Tignino, Raya Marina Stephan, Renée Martin-Nagle and Owen McIntyre

For more information about this series, please visit: www.routledge.com/series/WATER

Strengthening Cooperation over Transboundary Groundwater Resources

Edited by
Gabriel Eckstein and Alice Aureli

IWRA

Routledge
Taylor & Francis Group
LONDON AND NEW YORK

First published 2022
by Routledge
2 Park Square, Milton Park, Abingdon, Oxon OX14 4RN

and by Routledge
605 Third Avenue, New York, NY 10158

Routledge is an imprint of the Taylor & Francis Group, an informa business

British Library Cataloguing in Publication Data
A catalogue record for this book is available from the British Library

ISBN: 978-0-367-67314-7 (hbk)
ISBN: 978-0-367-67317-8 (pbk)
ISBN: 978-1-003-13077-2 (ebk)

DOI: 10.4324/9781003130772

Typeset in Times New Roman
by Newgen Publishing UK

Publisher's Note
The publisher accepts responsibility for any inconsistencies that may have arisen during the conversion of this book from journal articles to book chapters, namely the inclusion of journal terminology.

Contents

Citation Information

The following chapters were originally published in *Water International*, volume 36, issue 5 (2011). When citing this material, please use the original page numbering for each article, as follows:

Chapter 1
Strengthening cooperation on transboundary groundwater resources
Alice Aureli and Gabriel Eckstein
Water International, volume 36, issue 5 (2011), pp. 549–556

Chapter 2
Codification of the Law of Transboundary Aquifers (Groundwaters) by the United Nations
Chusei Yamada
Water International, volume 36, issue 5 (2011), pp. 557–565

Chapter 3
Managing buried treasure across frontiers: the international Law of Transboundary Aquifers
Gabriel E. Eckstein
Water International, volume 36, issue 5 (2011), pp. 573–583

Chapter 4
The customary law applicable to internationally shared groundwater
Joseph W. Dellapenna
Water International, volume 36, issue 5 (2011), pp. 584–594

Chapter 5
The International Law Commission's flawed Draft Articles on the Law of Transboundary Aquifers: the way forward
Stephen C. McCaffrey
Water International, volume 36, issue 5 (2011), pp. 566–572

Chapter 6
The World Bank policy and practice for projects affecting shared aquifers
Salman M.A. Salman
Water International, volume 36, issue 5 (2011), pp. 595–605

Chapter 7
Governance of transboundary aquifers: new challenges and new opportunities
Jamie Linton and David B. Brooks
Water International, volume 36, issue 5 (2011), pp. 606–618

Chapter 8
Interpreting the unknown: uncertainty and the management of transboundary groundwater
Anita Milman and Isha Ray
Water International, volume 36, issue 5 (2011), pp. 631–645

Chapter 9
Unitization: a lesson in collective action from the oil industry for aquifer governance
W. Todd Jarvis
Water International, volume 36, issue 5 (2011), pp. 619–630

Chapter 10
Transboundary groundwater law in Europe: a look at an evolving multi-level regime
Götz Reichert
Water International, volume 36, issue 5 (2011), pp. 686–691

Chapter 11
The Agreement on the Guarani Aquifer: a new paradigm for transboundary groundwater management?
Pilar Carolina Villar and Wagner Costa Ribeiro
Water International, volume 36, issue 5 (2011), pp. 646–660

Chapter 12
Keeping it legal: transboundary management challenges facing Brazil and the Guarani
David N. Cassuto and Romulo S.R. Sampaio
Water International, volume 36, issue 5 (2011), pp. 661–670

Chapter 13
Sustainable transboundary groundwater management under shifting political scenarios: the Ceylanpinar Aquifer and Turkey–Syria relations
Mehmet Öztan and Mark Axelrod
Water International, volume 36, issue 5 (2011), pp. 671–685

Notes on Contributors

Alice Aureli Chief of Section on Groundwater Systems and Water for Human Settlements, United Nations Educational, Scientific, and Cultural Organization, Paris, France.

Mark Axelrod James Madison College and Department of Fisheries & Wildlife, Michigan State University, East Lansing, Michigan, USA.

David B. Brooks formerly with Friends of the Earth Canada, Ottawa, Ontario, Canada (retired).

Pilar Carolina Villar Sea Institute, Universidade Federal de São Paulo, Cidade Universitaria, São Paulo, Brazil.

David N. Cassuto Pace Law School and Brazil-American Institute for Law & Environment (BAILE), White Plains, New York, USA.

Wagner Costa Ribeiro Department of Human Geography, School of Philosophy, Literature and Human Sciences, University of São Paulo, Cidade Universitária, São Paulo, Brazil.

Joseph W. Dellapenna Peking University School of Transnational Law and Wuhan University, China, and Rapporteur of the Berlin Rules on Water Resources.

Gabriel Eckstein Texas A&M University School of Law, International Water Resources Association, International Water Law Project, Fort Worth, Texas, USA.

W. Todd Jarvis Department of Geosciences and the Institute for Water and Watersheds, Oregon State University, Corvallis, Oregon, USA.

Jamie Linton University of Limoges, France.

Stephen C. McCaffrey University of the Pacific, McGeorge School of Law, Sacramento, California, USA and Special Rapporteur for the International Law Commission on the Law of the Non-Navigational Uses of International Watercourses.

Anita Milman Department of Environmental Conservation, University of Massachusetts, Amherst, Massachusetts, USA.

Mehmet Öztan Department of Geology and Geography, West Virginia University, USA.

Isha Ray Energy & Resources Group, University of California, Berkeley, California, USA.

Götz Reichert Centrum für Europäische Politik (CEP), Freiburg, Germany.

Salman M.A. Salman IWRA Member.

Romulo S.R. Sampaio Research Program on Law & the Environment, FGV Law School, Rio de Janeiro, Brazil and Brazil-American Institute for Law & Environment (BAILE), White Plains, New York, USA.

Chusei Yamada former Special Assistant to the Minister for Foreign Affairs, Government of Japan, former Special Rapporteur for the United Nations International Law Commission on shared natural resources (deceased).

Introduction: The ebb and flow of law and governance for transboundary aquifers and groundwater resources

Gabriel Eckstein and Alice Aureli

Over the past decade, since the articles in this compilation first appeared in 2011, much has taken place in the realm of transboundary groundwater resources. Novel research projects have been initiated, new data and information has been collected, and a number of new agreements and arrangements have been forged among states sharing transboundary groundwater resources. While there is much that remains to be done, a great deal has been achieved.

Today, reliance on groundwater globally continues to expand at alarming rates. At the global level, approximately 1,000 cubic kilometers of groundwater is extracted every year from aquifers to meet agricultural, industrial, and human needs (Famiglietti, 2014). Extracted at around two hundred times more than oil, by volume, groundwater continues its dominance as the most extracted natural resource on the planet (Eckstein, 2017). Around 45% of humanity's freshwater needs for everyday domestic uses, such as cooking and hygiene, come from groundwater (Margat and van der Gun, 2013), and more than half of all drinking water originates from groundwater (Mukherjee et.al., 2021).

While precise figures have yet to be computed, it is evident that a significant proportion of the world's population resides over aquifers that are directly transboundary or otherwise connected to transboundary watercourses. Most of the world's transboundary river basins, which contain more than forty percent of the world's population (UN Environmental Programme, 2002), have hydraulically connected aquifers. Moreover, numerous communities and nations around the world rely on cross-border fossil aquifers as their primary source of freshwater. Accordingly, it is evident that transboundary aquifers and groundwater resources are critical to a significant proportion of the world's population.

As a result, groundwater has remained firmly on the international agenda in the global quest to achieve water security and sustainability. In 2019, the interagency mechanism that coordinates UN and international organization efforts on water and sanitation issues, UN-Water, announced that the theme for World Water Day in 2022 will be "Groundwater: Making the Invisible Visible." Moreover, in the transboundary context, in December 2021, the UN Educational, Scientific and Cultural Organization (UNESCO) reprised its International Shared Aquifer Resources Management (ISARM) conference in conjunction with the International Association of Hydrogeologists, International Groundwater Resources

Assessment Centre, International Water Resources Association, and a host of other global partners.

In addition, more efforts are now being placed on exploring both the technical characteristics and governance aspects of specific transboundary aquifers. Under the Governance of Groundwater Resources in Transboundary Aquifers (GGRETA) framework, projects have been undertaken for: the Ocotepeque–Citalá Aquifer in El Salvador and Honduras; the Stampriet–Kalahari/Karoo Aquifer shared by Namibia, Botswana, and South Africa; and the Pretashkent Aquifer underlying Kazakhstan and Uzbekistan. Financed by the Swiss Agency for Development and Cooperation, the GGRETA project is implemented by the UNESCO Intergovernmental Hydrological Programme in close cooperation with teams of national experts for pilot projects on three transboundary aquifers. Previous phases of the project were also implemented in cooperation with the International Union for Conservation of Nature and the UNESCO International Groundwater Resources Assessment Centre, delivering a shared science-based understanding of the pilot projects, as well as tools for groundwater management, institutional strengthening, and transboundary dialogue (UNESCO-IHP Groundwater Portal, n.d.).

Furthermore, more nations are beginning to explore and engage with their neighbors over their cross-border groundwater resources. Building on the GGRETA initiative, in 2019, local, regional, and national institutions, communities, and stakeholders in El Salvador and Honduras entered into the *Ocotepeque–Citalá Statement of Intent* (SoI) in an effort to coordinate their management efforts across the international frontier. While aspirational in nature, the SoI proposes the development of a more formal binational strategy for the conservation, protection, and sustainable utilization of the aquifer, and calls on the Parties to share and harmonize data and information on the aquifer (Sindico, 2019). In a similar vein, in 2017, the Governments of Botswana, Namibia, and South Africa established a Multi-Country Cooperation Mechanism (MCCM) for the joint governance and management of the Stampriet Transboundary Aquifer System (STAS). Nested in the Orange-Senqu River Commission's Ground Water Hydrology Committee, the long-term vision of the MCCM is to become a permanent institution whose mission is to advise and strategize with the STAS countries on managing the aquifer and its resources.

Separately, in 2017, Belgium and France entered into the *Convention for the Provision and Exchange of Data Relating to the Management of Groundwater from Carboniferous Limestone*. As a binding instrument, the Convention focuses exclusively on the exchange of data and information about the aquifer, including its conditions and use, between the two aquifer riparians. Likewise, in 2015, Jordan and Saudi Arabia formalized the *Agreement between the Government of the Hashemite Kingdom of Jordan and the Government of the Kingdom of Saudi Arabia for the Management and Utilization of the Ground Waters in the Al-Sag/Al-Disi Layer*. That official instrument centers on restricting groundwater extraction and protecting groundwater quality in the binationally shared Al-Sag/Al-Disi Aquifer.

Irrespective of these initiatives, other efforts in the realm of transboundary aquifers and groundwater resources have been lackluster. There is still no broad consensus on the international principles or procedures by which nations should manage their transboundary groundwaters. For example, the UN Draft Articles on the Law of Transboundary Aquifers (Draft Articles), which were introduced at the UN General Assembly in 2008, are still being considered by that august body. Although they were commended to the attention of the Member States, the Draft Articles were debated at the General Assembly in 2011, 2013,

2016, and 2019, and are again scheduled for the General Assembly's agenda in 2022 (UN General Assembly, 2019). While the continued deliberation over these articles may appear as a hopeful sign, the fate of those Draft Articles remains unclear.

Moreover, in comparison with surface water, groundwater is still treated with less urgency and importance on the international agenda. For example, when establishing the framework of indicators for the Sustainable Development Goals (SDG), no indicator was explicitly identified to assess groundwater despite the reality that groundwater resources and aquifers are relevant for many of the SDG indicators and were included in the scope of their methodology as elements to be assessed (e.g., SDG 6.3 water quality and SDG 6.6 water-related ecosystems). As a result, many countries report very limited information on their transboundary aquifers and groundwater resources and only in the context of SDG 6.5.2 on transboundary cooperation. As revealed in the 2nd reporting effort for this indicator, which occurred in 2020, more needs to be done to build the capacity of nations to develop, assess, and improve data about their shared groundwater resources (UN and UNESCO, 2021). Without a more concerted and targeted effort, the critical importance of groundwater, and specifically in a transboundary context, will remain overlooked and underestimated.

With this in mind, it is clear that the compilation of outstanding articles in this book remain as relevant as they were when first published. In reissuing them in this format, we hope to provide water professionals and the public at large with insight into the management, development, and conservation of transboundary aquifers and groundwater resources, and offer decision-makers ideas, examples, and recommendations that will broaden opportunities for cooperation over these critical freshwater resources. As we noted in our original introduction, "Water is life's most important natural resource. And groundwater is humanity's most vital supply of that life-giving substance." That reality will likely continue in perpetuity.

References

Agreements and Arrangements

Agreement between the Government of the Hashemite Kingdom of Jordan and the Government of the Kingdom of Saudi Arabia for the Management and Utilization of the Ground Waters in the Al-Sag/Al-Disi Layer (2015). Jordan and Saudi Arabia.

Convention for the Provision and Exchange of Data Relating to the Management of Groundwater from Carboniferous Limestone (2017). Belgium and France.

Ocotepeque–Citalá Statement of Intent (2019). El Salvador and Honduras.

Articles, Books, and Reports

Eckstein, G. (2017). *The International Law of Transboundary Groundwater Resources*. Routledge.

Famiglietti, J.S. (2014). The Global Groundwater Crisis. *Nat. Clim. Change* 4 (11), pp. 945–948.

Margat, J. and van der Gun, J. (2013*) Groundwater Around the World: A Geographic Synopsis*. CRC Press.

Mukherjee A. et.al. (2021). Global groundwater: from scarcity to security through sustainability and solutions, in Mukherjee A. et.al. (eds.), *Global Groundwater: Source, Scarcity, Sustainability, Security, and Solutions*. Elsevier, pp. 3–20.

Orange-Senqu River Commission (2018). Operationalization of the Stampriet Transboundary Aquifer System (STAS) Multi-Country Cooperation Mechanism (MCCM) – Terms of Reference, https://groundwaterportal.net/sites/default/files/TORs_UNESCO_ORASECOM_Final.pdf.

Sindico, F. (2019). The Ocotepeque – Citalá Statement of Intent: A first step towards transboundary aquifer cooperation in Central America?, University of Strathclyde Centre for Environmental Law and Governance SCELG Dialogue.

UN and UNESCO (2021). Progress on Transboundary Water Cooperation - Global status on SDG indicator 6.5.2 and acceleration needs, [Website forthcoming].

UN Environmental Programme (2002). Atlas of International Freshwater Agreements.

UN General Assembly (2019). Resolution adopted by the General Assembly: The law of transboundary aquifers. *A/Res/74/193*, 18 December 2019.

UNESCO-IHP Groundwater Portal (n.d.). Groundwater Resources Governance in Transboundary Aquifers (GGRETA), https://groundwaterportal.net/ggreta.

Strengthening cooperation on transboundary groundwater resources

Alice Aureli and Gabriel Eckstein

Water is life's most important natural resource, and groundwater is humanity's most vital source of that life-giving substance. Global freshwater resources contained in aquifer systems are two orders of magnitude greater than those found in rivers, lakes and other surface freshwaters combined. Moreover, more than one half of the world's population today is dependent on groundwater for its basic needs (Eckstein 2003).

A large number of the aquifers on which humanity depends lay across and underneath international political borders. According to the most recent studies of the United Nations Educational, Scientific and Cultural Organization (UNESCO), transboundary aquifers underlie the territory of nearly every non-island nation (Puri and Aureli 2009). The International Shared Aquifer Resources Management (ISARM) initiative, which was launched by UNESCO's International Hydrological Programme (IHP) in 2000, has identified more than 270 transboundary aquifers worldwide (Stephan 2009).

In recent years, as communities and nations realized the significance of these shared freshwater resources, transboundary aquifers have become the subject of growing attention among aquifer riparians and the international community. Questions are now being raised with regard to the mechanisms for governing shared groundwater resources, the rights that aquifer riparians enjoy from a transboundary aquifer and the reponsibilities that these nations might owe to other aquifer riparians.

As a result, in 2002, building on its prior work that led to the formulation of the 1997 UN Convention on the Law of the Non-navigational Uses of International Watercourses, the United Nations International Law Commission (UNILC) decided to continue its work on "shared natural resources" focusing on transboundary aquifers (Yamada 2003). As part of that effort, UNESCO-IHP, through its ISARM initiative, mobilized hydrogeologists, water resources managers, water law experts and groundwater administrators to provide advisory support to the UNILC. Within a relatively short period of six years, the UNILC completed 19 draft articles, which it submitted in 2008 to the 63rd Session of the United Nations General Assembly (UNGA). Thereafter, the UNGA adopted the Resolution on the Law of Transboundary Aquifers (A/RES/63/124) in December 2008, which included in its annex the 19 draft articles. The draft articles have now been placed on the provisional agenda of the 66th session of the UNGA, scheduled for Autumn 2011.

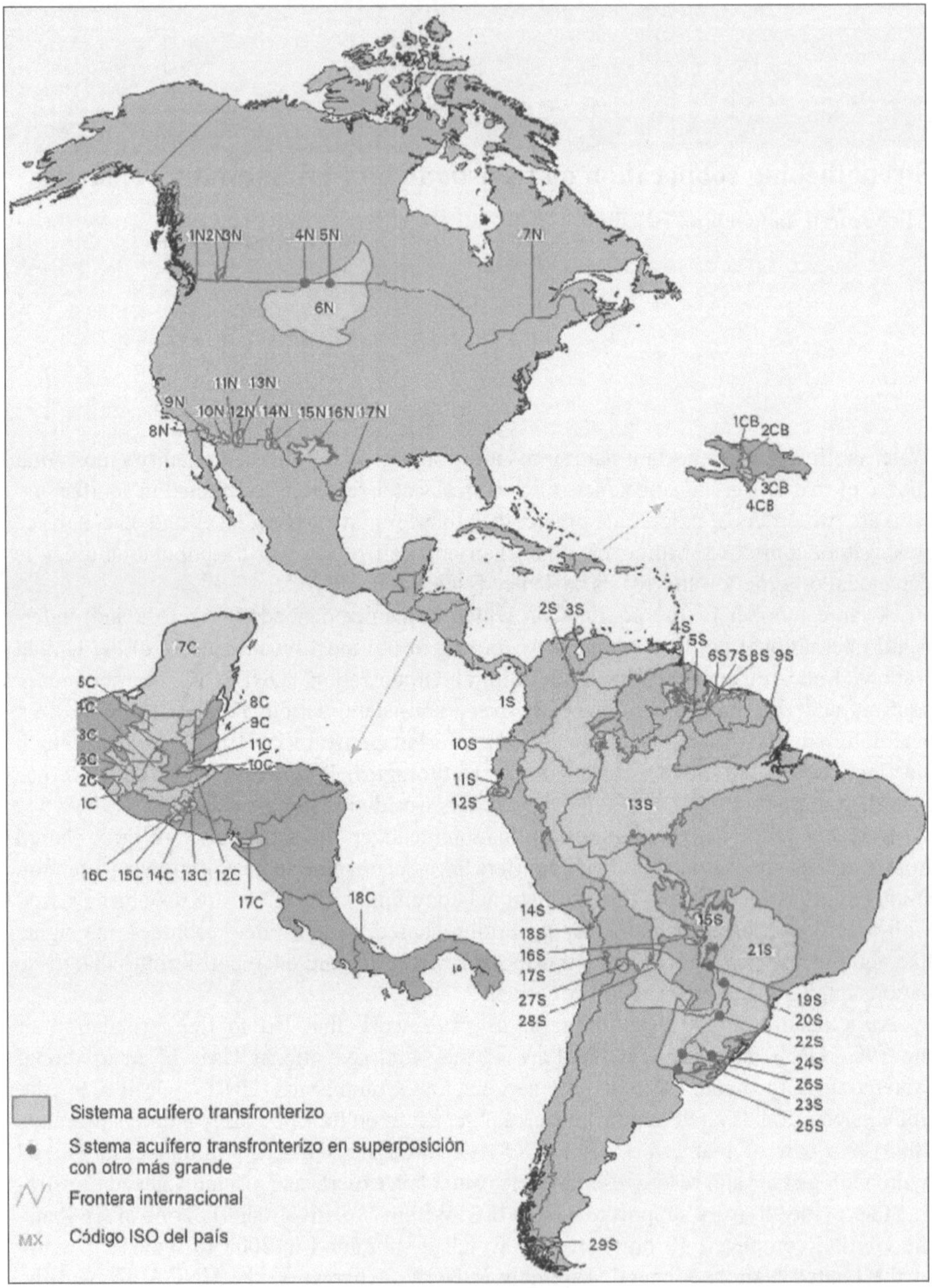

Figure 1. Transboundary aquifers in Latin America and the Caribbean (UNESCO/OAS 2007).

The issues related to the governance of transboundary aquifers have yet to be fully developed. Nevertheless, the law and management of these critical resources are slowly evolving as nations continue to explore their shared aquifers. As a consequence, there are an increasing number of examples of good cooperation among countries sharing transboundary aquifers.

As a means of furthering these developments and encouraging cooperation among aquifer riparians, the International Water Resources Association and UNESCO-IHP are delighted to have teamed up to prepare this special issue *of Water International* on *Strengthening cooperation on transboundary groundwater resources.* This collaboration provides both organizations a unique opportunity to present outstanding scholarship in this increasingly important field as well as provide decision makers with ideas and examples that will broaden and advance the conversation toward more effective cooperation.

UNESCO is especially poised to address transboundary aquifer issues. In particular, UNESCO-IHP is the only intergovernmental scientific programme of the United Nations (UN) system devoted to hydrology and hydrogeology, water resources management and water education. Since its inception in the late 1970s, UNESCO-IHP has paid due attention to studies on groundwater resources and aquifer characteristics, enhancing the role that aquifers play in supporting human activities and ecosystems.

In its current 7th Phase (2008–13), UNESCO-IHP has intensified the study of groundwater resources on a global scale and is devoting particular attention to the development of guidelines and frameworks for the protection and wise use of aquifers around the world. In particular, UNESCO-IHP has emphasized that awareness of groundwater resources and aquifer systems still remains limited in many regions of the world. Lack of attention can have more serious consequences when the transboundary nature of these systems is not identified.

Until recently there was very little knowledge about transboundary aquifers and regional and global estimations were not available. The Intergovernmental Council of UNESCO-IHP has made the issue of transboundary aquifers a priority and member states have tasked UNESCO-IHP to lead a global effort to fill this knowledge gap. UNESCO is now in the process of preparing a global assessment of transboundary aquifers, a review of their relevance to the overall global water balance and the formulation of a set of recommendations for the sustainable management of these systems. Consequently, in June 2000, at the 14th Session of the UNESCO-IHP Intergovernmental Council, member states adopted Resolution XIV-12 launching the ISARM initiative. Over the past decade, numerous organizations have provided ISARM with their support and cooperation, including the United Nations Food and Agriculture Organization (FAO), United Nations Economic Commission for Europe (UNECE), United Nations Economic and Social Commission for Western Asia (UNESCWA), Organization of American States (OAS) and International Association of Hydrogeologists (IAH).

The ISARM initiative is based on a multidisciplinary approach and the effective cooperation between various disciplines. It includes the evaluation of the hydrogeological, environmental, legal, institutional and socio-economic status of transboundary aquifer systems and the development of an inventory that includes information on:

- location, extent and type of transboundary aquifers
- quantity and quality of transboundary aquifers
- functioning and dynamics of aquifers
- groundwater-dependent ecosystems
- existence of aquifer-monitoring activities
- potential aquifer pollution sources
- uses, problems and trends in the state of groundwater resources
- socio-economic relevance of aquifers
- institutional aspects and management of groundwater resources at national, regional and sub-regional levels
- legal and administrative mechanisms for transboundary aquifers
- existing bilateral or regional agreements for transboundary aquifers.

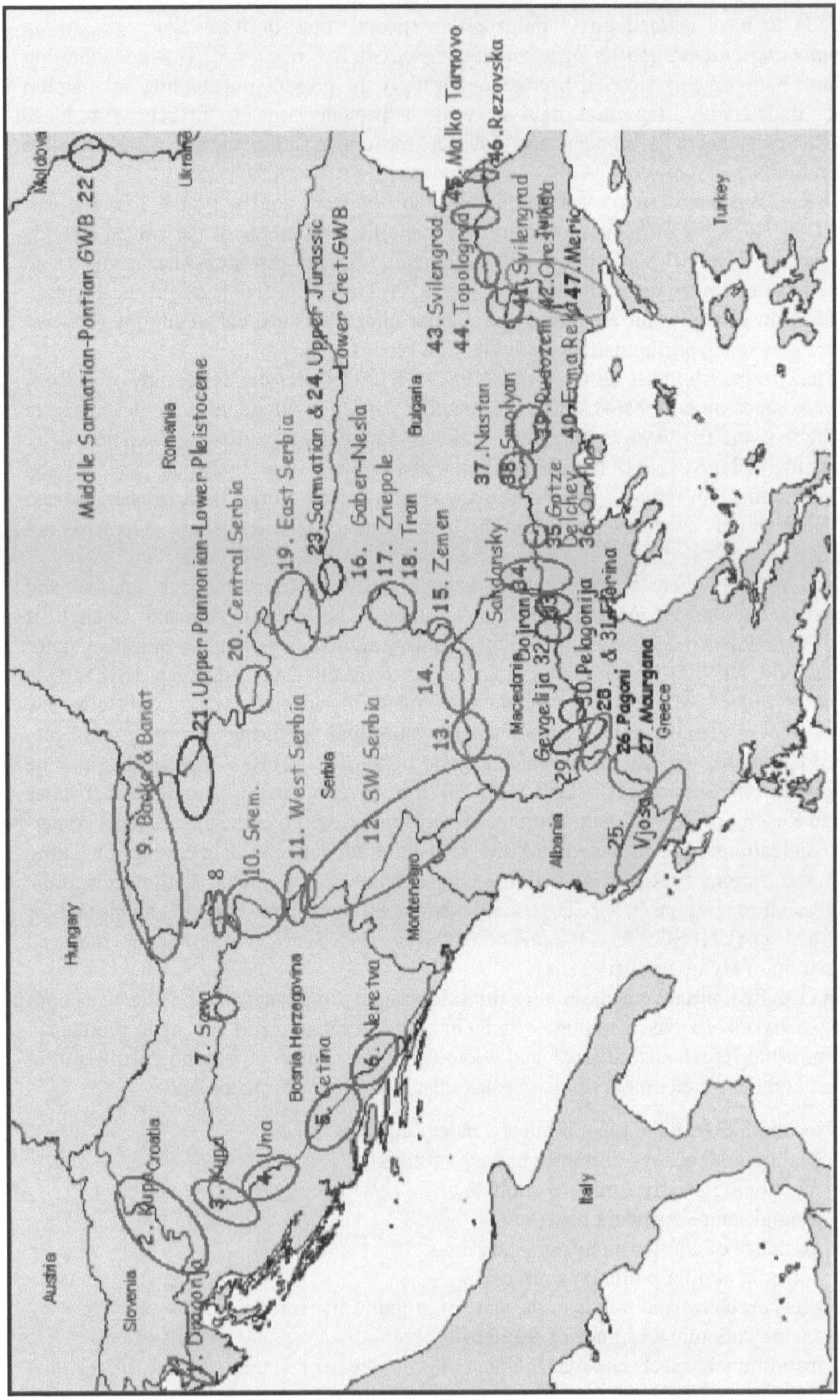

Figure 2. Transboundary aquifers in Southeast Europe (UNESCO 2008a).

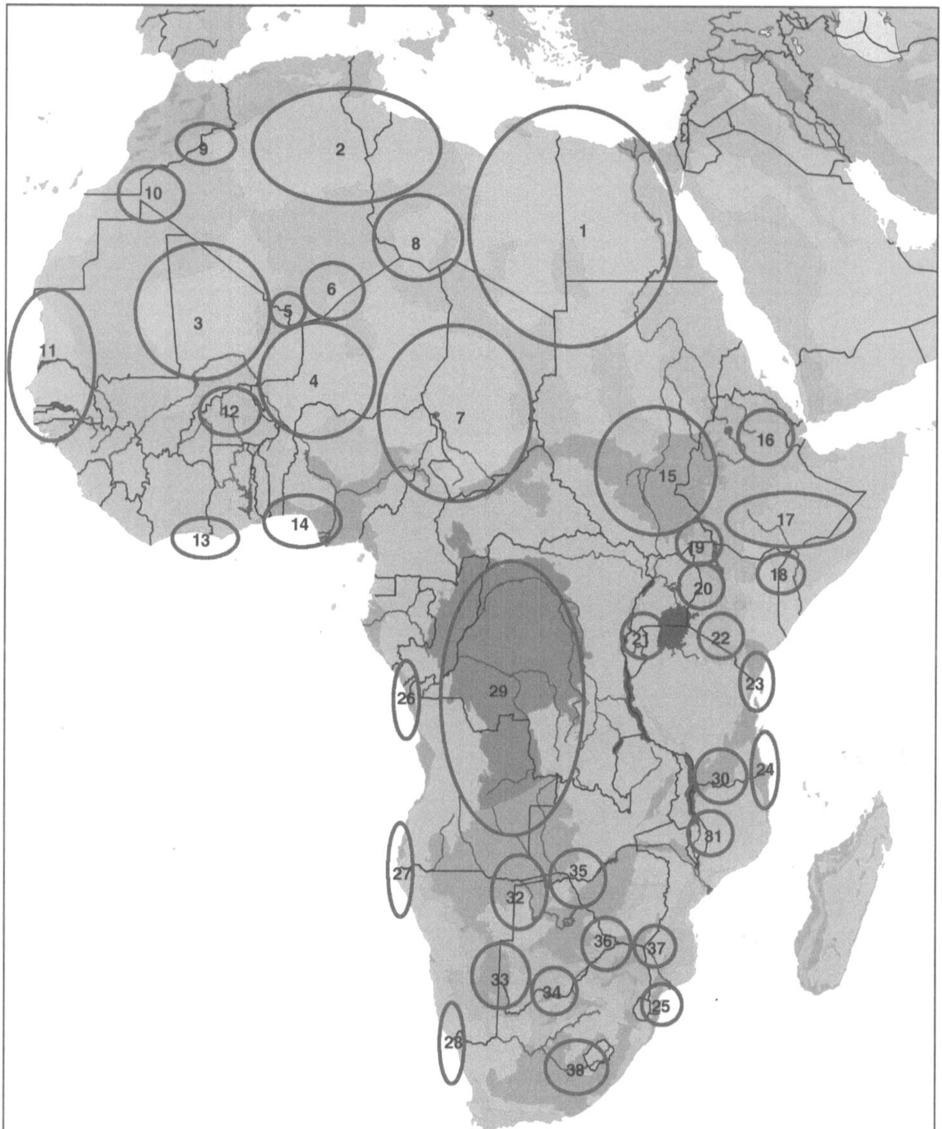

Figure 3. Transboundary aquifers in Africa (UNESCO 2004).

The objectives for carrying out this inventory are to enhance the knowledge base on aquifers globally and to build trust among partners, scientists and experts from both sides of every transboundary aquifer boundary around the world. The study of transboundary aquifers necessitates cross-border cooperation, and the agreement among scientists and other professionals is an essential precursor to the sustainable governance of shared aquifers.

With the support of numerous governments and international organizations, ISARM published the *Atlas of transboundary aquifers: global maps, regional cooperation and local inventories* in 2009.[1] This inventory includes information on 273 transboundary aquifers

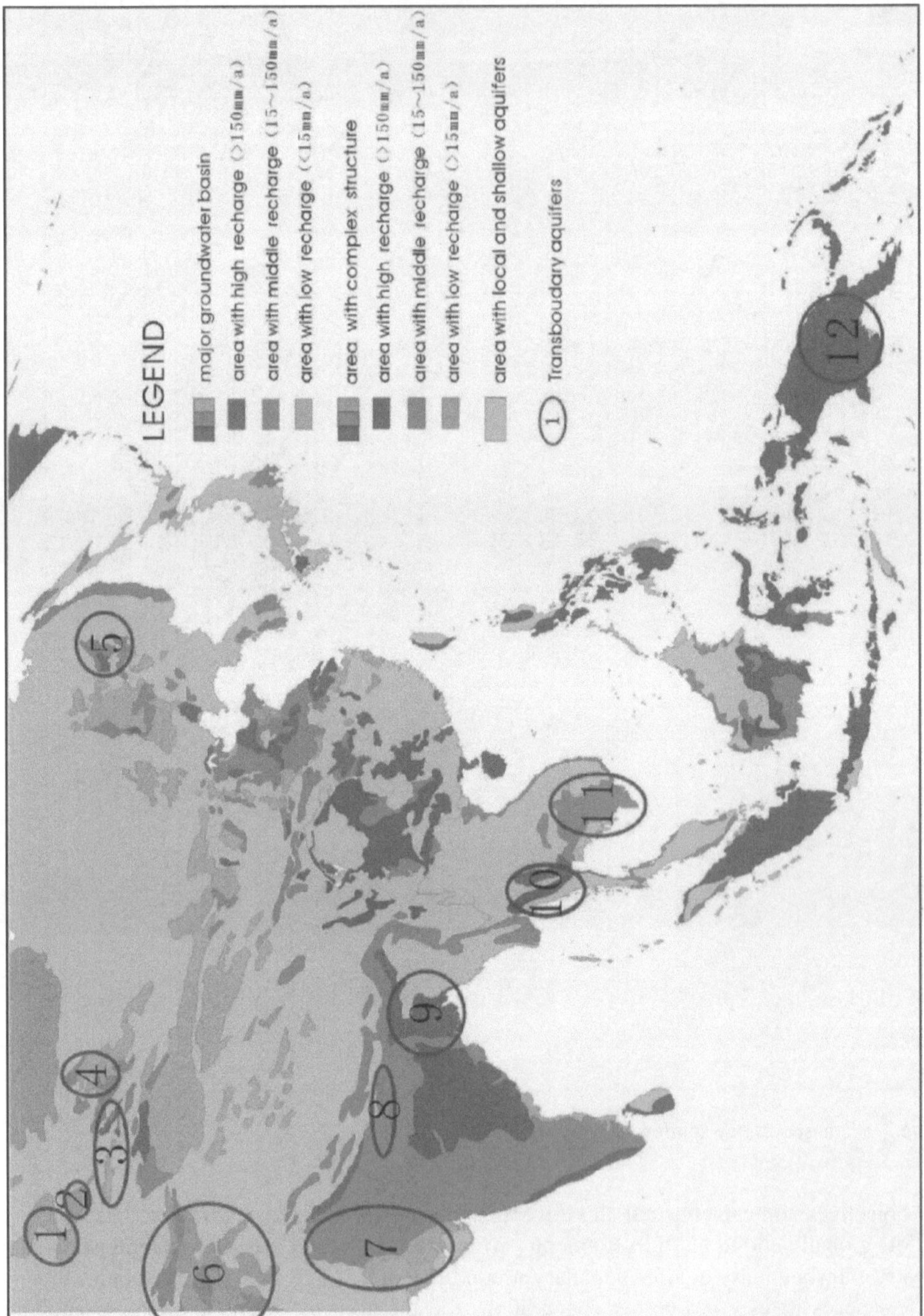

Figure 4. Transboundary aquifers in Asia (UNESCO 2008b).

around the world and revealed that the number of transboundary aquifers already identified is comparable to that of international watercourses (Stephan 2009). Regional inventories have also been completed for Latin America and the Caribbean (Figure 1), Southeast Europe (Figure 2), and in some sub-regions of Africa (Figure 3). However, the volume of data on the transboundary aquifers identified to date will likely increase in future years as a result of ISARM's ongoing investigations, in particular in Asia (Figure 4) and Africa.

Building on its ISARM initiative, UNESCO-IHP launched the World-wide Hydrogeological Mapping and Assessment Programme (WHYMAP)[2] in 2002 to identify, map and categorize all of the world's transboundary aquifers. While the assessment programme is still underway, it has already classified more than 155 transboundary aquifers on the European continent, 38 in Africa, 68 in the Americas and 12 in Asia (Stephan 2009). In comparison, there are 263 watercourses crossing international boundaries (UNEP 2002). In addition, UNESCO-IHP has prepared the *Groundwater resources of the world*, and published the first worldwide map of transboundary aquifers.[3] As a result, UNESCO-IHP has provided an invaluable contribution to better defining the characteristics of the world's groundwater resources on a global scale, and to the formulation of strategies for the protection and sustainable management of shared aquifer resources.

ISARM's first 10-year phase, which recently concluded, greatly enhanced existing knowledge of and fostered increased cooperation over transboundary groundwater resources throughout the world. ISARM's second phase, initiated in 2011, is designed to better define hydrogeological conditions at the local and sub-regional level as well as identify possible knowledge gaps which may have an impact on effective groundwater resource management in specific areas. It will consider the wide and still challenging task of initiating activities on many of the inventoried aquifers as well as address the limitations in terms of manpower, institutional and funding resources in many regions that enable a well-focused and pragmatic programme. Actions on specific aquifers are urgently needed and one of the main partners undertaking this task is the Global Environment Facility under its International Waters Focal Area. Future activities of ISARM will also be aimed at supporting the implementation of the UNGA Resolution and continuing to provide scientific input to the formulation of bilateral, regional and international protocols and conventions.

As an important component of this second ISARM phase, this special issue of *Water International* is intended to provide water professionals, decision-makers, and the public at large with insight into various developments and case studies from around the world related to cooperation over transboundary aquifers. While the immediate goal of this UNESCO-IHP/*Water International* partnership is the provision of relevant and interesting information, it is hoped that this effort also generates additional attention to and discussions about transboundary groundwater resources and, ultimately, furthers collaborative efforts globally.

Notes

1. The *Atlas* is available at http://www.isarm.net/publications/324.
2. WHYMAP is available at http://www.whymap.org/.
3. The WHYMAP map *Groundwater resources of the world – transboundary aquifer systems* is available at http://www.whymap.org/whymap/EN/Downloads/Global_maps/spec_ed_2_map_pdf.pdf?__blob=publicationFile&v=2

References

Eckstein, G., 2003. A hydrogeological approach to transboundary ground water resources and international law. *American University International Law Review*, 19 (2), 201–258.

Puri, S. and Aureli, A., eds. 2009. *Atlas of transboundary aquifers. Global maps, regional cooperation and local inventories*. Paris: UNESCO-IHP ISARM Programme.

Stephan, R.M., ed., 2009. *Transboundary aquifers: managing a vital resource. The UNILC Draft Articles on the Law of Transboundary Aquifers*. Paris: UNESCO.

UNESCO, 2004. ISARM-Africa. *Managing shared aquifer resources in Africa*. B. Appelgren, ed. ISARM-Africa. IHP-VI, series on groundwater 8. Paris: UNESCO.

UNESCO, 2008a. *Internationally shared aquifers in the Balkan region: preliminary assessment*. Paris: UNESCO/ INWEB.

UNESCO, 2008b. *Transboundary aquifers in Asia with special emphasis to China*. Han Zaisheng, Wang Hao and Chai Rui associated with R. Jayakumar, Liu Ke and Wang Jin, eds. Report of ISARM Asia pilot case study. Paris: UNESCO.

UNESCO/OAS, 2007. *Sistemas acuíferos transfronterizos en las Américas. Evaluación Preliminar*. Programa UNESCO/OEA ISARM Américas, Serie ISARM Américas N°1. Montevideo/Washington DC: UNESCO-IHP/OAS.

UNEP, 2002. *Atlas of international freshwater agreements*. Nairobi: United Nations Environment Programme.

Yamada, C., 2003. *Shared natural resources: first report on outlines*, UN Doc. A/CN.4/533/ ¶ 21 (2003). Available from: http://internationalwaterlaw.org/bibliography/UN/UNILC/Groundwater/Yamada%201st%20Report-Addendum.pdf [Accessed 25 August 2011]

Codification of the Law of Transboundary Aquifers (Groundwaters) by the United Nations

Chusei Yamada

Having served as a Special Rapporteur for the United Nations International Law Commission's topic of Shared Natural Resources, the author was responsible for formulating the Draft Articles on the Law of Transboundary Aquifers, which was submitted to the United Nations General Assembly in 2008. In this article, the author discusses the background to the Draft Articles, their formulation in cooperation with hydrogeologists and groundwater administrators, the principles and concepts contained in the Draft Articles, positions of various states, and the future prospect of the Draft Articles.

Introduction

I served as a member of the United Nations (UN) International Law Commission (ILC) from Japan for 17 years from 1992 to 2009. While on the ILC, I served as a Special Rapporteur for the ILC's topic of Shared Natural Resources and was responsible for formulating the Draft Articles of the Law on Transboundary Aquifers (groundwaters), which was submitted to the UN General Assembly in 2008. I would like to present the background to these Draft Articles, how they were formulated in cooperation with hydrogeologists and groundwater administrators, main elements of the principles and concepts contained in the Draft Articles, positions of various states and the future prospect of the draft articles.

International law

I would like to emphasize first the importance of the codification of international law. In order to secure justice and order and to settle any dispute among states by peaceful means, it is essential to establish the "rule of law" among the world community. The main sources of international law are treaties and customary international law. Treaties are like contracts that bind only those states that become parties to those instruments. States are fully acquainted with the rights and duties under treaties as they are expected to carefully examine them before they decide to become parties. On the other hand, customary international law binds all states of the international community regardless of their participation in the formulation of such customary norms. Customary international law is defined as "international custom, as evidence of a general practice accepted as law" (Article 38, Statute of the International Court of Justice).

Historically speaking, in the absence of a world legislature, international law has largely developed as customary law. However, it is often difficult to ascertain precisely what the customary rules are, and states have differences in the interpretation of those rules. Furthermore there exist many lacunae in customary international law. These factors present serious difficulties to practitioners of international law. In order to remove the ambiguities of customary international law, there have been efforts at the restatement of customary rules to be agreed upon by states. This process is referred to as the codification of international law. The Charter of the UN, in Article 13, 1(a), provides that one of the important functions of the UN General Assembly is to promote the codification of international law.

The ILC was established in 1947 as a subsidiary organ of the UN General Assembly whose mandate is to prepare basic documents for the codification of international law in the form of draft articles. Article 13, 1(a) of the UN Charter contains two concepts. The first is the "codification of international law" and the other is "progressive development of international law". The Statute of the ILC defines them separately in its Article 15. However, the ILC has found it difficult to separate one from the other in practice. Its codification now contains both concepts, with increasing focus in recent years on the progressive development of international law. The UN has so far adopted 13 codification treaties on the basis of the work of the ILC, including such important conventions as Diplomatic and Consular Relations, Law of Treaties, Law of the Sea and Jurisdictional Immunity of States and Their Properties.

Why water?

Let me now address the question of "water". With respect to freshwaters, the Rhine and Danube rivers in Europe were subject to international regulations as early as the beginning of the nineteenth century for the purpose of free navigation. In fact, their river commissions, which were established for regulating international navigation, were the precursors of present international administrative organizations. The first time the UN dealt with transboundary freshwater resources occurred when it instructed the ILC in 1970 to take up the study of the law of the non-navigational uses of international watercourses.

Since the mid-twentieth century, large projects were undertaken along the international rivers of the world, notably the construction of dams and other facilities for the purposes of drinking, electric-power generation, irrigation and other uses. These upstream activities inevitably have downstream effects, and in many cases, threaten to cause significant adverse effects upon downstream states. To regulate these activities, the UN adopted in 1997 the Convention on the Law of the Non-Navigational Uses of International Watercourses (hereafter referred to as the 1997 International Watercourses Convention) on the basis of work by the ILC. While that convention theoretically covers groundwater resources as they are physically linked to international surface waters, it essentially was meant to regulate surface waters. In the preparation of that convention, the ILC did discuss the question of whether to include groundwater within the project. Although it recognized the need to properly deal with groundwater, the ILC decided that a separate study was required for that purpose.

Groundwater and the ILC

The UN soon became aware of the rapidly expanding exploitation of groundwater resources for potable, industrial and irrigation uses in both developed and developing countries and of the resulting critical overexploitation and pollution problems. Hence it instructed the ILC in 2001 to proceed with the work on "Shared Natural Resources" which were generally understood to include groundwater as well as oil and natural gas. The ILC embarked on the work in 2002, appointing me as its Special Rapporteur for this new topic. Though there exist many similarities between groundwater and oil and natural gas, there is also much dissimilarity between them. Upon my recommendation, the ILC chose to adopt a step-by-step approach by embarking first on the work on transboundary groundwater resources as a follow up to the 1997 International Watercourses Convention.

The codification work on the law of transboundary groundwater required a multi-disciplinary process. However, the ILC consists of 34 members with a recognized competence in international law. As such, it does not possess scientific and technical knowledge of groundwater or expertise for proper management of such resources. Since the United Nations Educational, Scientific and Cultural Organization (UNESCO) has great experience coordinating global water issue for various UN organizations, it mobilized a team of hydrogeologists (groundwater scientists), groundwater administrators and water lawyers to assist me and the ILC. Without their untiring and valuable support, the ILC would not have been able to formulate the Draft Articles on the Law of Transboundary Aquifers. For example, during the work with these outside experts, it became clear that international lawyers often employ different concepts and terminologies from those used by scientific professionals. Accordingly, great effort was needed to select terms and expressions to be used in the Draft Articles that could be commonly understood by scientists, administrators and lawyers as well as by lay people.

About groundwater

The formulation of the Draft Articles by the ILC had to be based on firm scientific and technical evidence. Less than 1% of water on earth is readily accessible fresh water. Of this, 97% is located underground in aquifers. These aquifers exist in every continent on earth.

Rapid expansion in the exploitation of groundwater has been taking place since the 1950s in industrialized states and since the 1970s in developing states. Currently, 50% of potable water, 40% of industrial water and 20% of irrigation water is supplied by groundwater. Groundwater is the single most extracted raw material in the world. How groundwater is used varies in different regions. For example, it is used in the Russian Federation mainly for domestic purposes, in India for agriculture and in Central Asia for industries. Through the process of soil filtration, groundwater is usually of high quality and contains various nutritious minerals. Hence, bottled mineral water sold worldwide primarily comes from groundwater resources, although some is artificially produced.

The science of surface water is very old. You might have visited Roman aqueducts, most of which are located in southern Europe from Spain to Turkey. These aqueducts were built from the third century BC to the third century AD to transport water from their sources to cities. Building these aqueducts was possible because the science of surface waters – hydrology – was already highly developed then. Although humankind has tapped groundwater since time immemorial, the science of groundwater, especially of its flow, is a much newer field, dating back only 150 years to when Henry Darcy carried out various hydrologic experiments in Dijon, France. Among his many contributions, Darcy explained the phenomenon of artesian flow and described water flow through aquifers in mathematical terms.

Groundwater resources are distributed widely in the world. Many aquifers are shared by neighbouring states. In western and southeastern Europe, 154 transboundary aquifers have so far been identified by UNESCO and the International Groundwater Resources

Assessment Center. Likewise, transboundary aquifers have been categorized in North America (17), Central America (18), the Caribbean (four), South America (29), Africa (41) and Asia and the Middle East (55).[1]

In South America, the now well-known Guarani Aquifer System underlies the territories of Argentina, Brazil, Paraguay and Uruguay. As this region receives a lot of rainfall, the aquifer receives significant recharge. In contrast, the Nubian Sandstone Aquifer System, which underlies the territories of Chad, Egypt, Libya and Sudan, is a non-recharging or fossil aquifer that is located in an arid region where it no longer receives meaningful recharge from rainfall.

An example of the current grave situation of aquifers in the world is worth mentioning. In the past, India was not self-sufficient in wheat and depended upon US aid supply based on US Public Law No. 480 which authorizes assistance to developing countries with surplus wheat. But in recent years, wheat production in the three northwestern states of India – Punjab, Haryana and Rajasthan – increased rapidly through the utilization of aquifers, allowing India to become self-sufficient in wheat for the first time since its independence in 1947. However, many Indian farmers are now in trouble because of the lowering of the water table and are having problems in financing deeper wells. Analyses conducted by the gravity observation satellite "Grace" revealed that 109 cubic kilometres of water, equivalent to four times the water volume of Lac Leman, were extracted over the last six years in these three Indian states, and that the aquifer there is in a very precarious condition.

Let us now examine the mechanism of the global hydrological cycle. When the rain falls on the surface of the earth, some of the rainwater evaporates back to the atmosphere; some runs off the surface, joins rivers, flows toward oceans and then evaporates back to the atmosphere; some is absorbed by plants; some infiltrates the soil and becomes groundwater. This last description, however, is not entirely precise. In a temperate zone with significant rainfall, as in Japan, the soil near the surface is usually wet. But the water in the shallow soil is not necessarily groundwater. Though it might be technically possible to separate this water from the soil, it would be expensive and would not be an economically viable process for obtaining fresh water.

So what is groundwater? Groundwater is the water found in the subsurface within a saturated geologic formation, also known as an aquifer. This means that underground water that is not inside of an aquifer is not considered groundwater. It also means that an aquifer is a matrix, consisting of such materials as sand, gravel, rock and other porous material, containing groundwater and through which groundwater flows.

Generally, there are two kinds of aquifers. A confined aquifer is a saturated geological formation that is both underlain and overlain by less permeable material. The water contained in the saturated zone of a confined formation is under pressure that is greater than atmospheric pressure. When a well is drilled and reaches this geological formation, the pressure causes the water to shoot up. This is one of the phenomena that Henry Darcy explained through his experiments, and that can also be observed in cases of oil and natural gas. In comparison, an unconfined aquifer is a saturated geological formation that is underlain by a less permeable layer, but is not overlain by a similar confining, less permeable material. The water contained in this geological formation is under pressure equal to that of atmospheric pressure. Therefore, when a well reaches this type of geologic formation, the water contained within does not shoot up but can be pumped out mechanically. In order to manage groundwater properly, we must regulate both the water and the geological formation containing such water. Accordingly, from now on, I will use the term employed by the ILC of "aquifer", which the ILC interpreted to encompass both the geologic formation and the water contained within it.

Threats to aquifers

What activities affect aquifers? First, there is the utilization of groundwater. Over-extraction can lower the water table in an aquifer and deplete the resource and can also destroy the geological formation. Second, activities other than the utilization of water in the surface area overlaying an aquifer, such as agriculture, the use of chemical fertilizers and pesticides, drainage and waste dumping, can pollute the aquifer. In a river, pollutants may be washed away by the flow of the river. Once an aquifer is polluted, pollutants stay there, as the rate of water flow within most aquifers is considerably slower than water flow in rivers. And it is extremely difficult to clean it up. In addition, paving and construction works on the surface and underground can block the recharge and discharge processes of an aquifer or destroy the geological formation itself. For example, the city of Kyoto, Japan overlays a huge aquifer. This aquifer has supported the city, Japan's old capital, for 1000 years. In the 1960s, residents came to notice the lowering of the water table in their wells in some parts of the city and had to dig deeper to extract water. That situation coincided with the construction of a subway network. It is believed that the subway construction blocked the aquifer's recharge process and destroyed parts of the geological formation. In the case of transboundary aquifers, these activities in one state may adversely affect neighbouring states that share the same aquifers, by reducing their share of the water or by polluting the resource.

As stated above, there are recharging aquifers and non-recharging aquifers. Non-recharging aquifers are found primarily in arid regions of the world. One of the most representative is the Nubian Sandstone Aquifer System in North Africa. The region of North Africa used to be a more temperate zone many thousands of years ago. The water stored in the aquifer today is the rainfall of those times. Radioactive tracers have confirmed that the system no longer receives any meaningful recharge. Since the 1960s, radioactive materials of cesium and tritium emitted from nuclear tests and krypton from nuclear industries have been floating in the atmosphere. If these aquifers had received recharge from rainfall during the last 50 years, radioactive traces would have been found in the aquifers. None have been found in the Nubian Sandstone Aquifer System. Therefore, any exploitation of non-recharging aquifers, such as this system, will result in depletion, just as in the case of mineral resources or oil and natural gas.

The ILC and transboundary aquifers

During its consideration of the topic, the ILC was provided by the experts with these valuable scientific and technical characteristics and factors of aquifers. It also found ample state practices and almost 400 relevant treaties, general, regional and bilateral, on the basis of which customary rules could be identified.[2] States have also shown keen interest in the ILC's work as aquifers exist in almost all states and the overwhelming majority of states share transboundary aquifers with their neighbouring states. Many of these states also transmitted their valuable inputs and observations to the ILC. Taking into account the advice of experts and observations from governments, the ILC formulated a final set of 19 Draft Articles on the Law of Transboundary Aquifers in 2008, the text of which was appended to UN General Assembly resolution A/RES/63/124 of 2008. The ILC also drafted detailed commentaries to each of the draft articles in Chapter IV of the ILC Report of 2008 (UN Document A/63/10).[3]

The ILC took 24 years to complete the Draft Articles on the Law of Non-Navigational uses of International Watercourses. That it was able to complete the codification work on transboundary aquifers in six years reflects its awareness of the critical situation of aquifers and of the urgent need to establish a legal framework for proper management of transboundary aquifers in order to achieve the objectives of equitable and reasonable utilization, protection of the environment and international cooperation. Some salient points of the Draft Articles are as follows:

Article 1: scope

The scope of the application of the Draft Articles is: (a) utilization of transboundary aquifer; (b) other activities that have or are likely to have an impact upon such aquifers; and (c) measures for the protection, preservation and management of such aquifers. The concept contained in paragraph (b) was not included in the 1997 International Watercourses Convention. As noted above, activities other than direct utilization, which are conducted above the aquifers, must be regulated. This point is also relevant to Article 6.

Article 2: use of terms

Article 2 provides definitions for terms used in the Draft Articles. "Aquifer" means both a geological formation which serves as a container and the water contained in the saturated zone of the formation. It is necessary to include the geological formation in the definition of aquifer in order to preserve the proper functioning of aquifer. It is also necessary to include the geological formation in order to regulate its own utilization such as for storage, disposal of waste, or the new experimental technique for carbon sequestration, used to store excess carbon from extracted oil and natural gas in geological formations. A definition of recharging aquifer is also provided in subparagraph (f). The threshold of negligibility is presented here because even non-recharging aquifers can receive some minimal amounts of recharge.

General principles

Articles 3 to 9 of Part Two set out general principles such as sovereignty, reasonable and equitable utilization, obligation not to cause significant harm and international cooperation.

Article 3: sovereignty of aquifer states

This article provides that each state has sovereignty over the portion of a transboundary aquifer located within its territory. I received critical comments from some international lawyers for the inclusion of this sovereignty clause, which might in their view diminish the value of the whole exercise. I share their apprehension, but we must squarely face the current state of affairs. It was the UN that passed the resolution "Permanent Sovereignty Over Natural Resources" (1803[XVII]) in 1962. Many aquifer states hold the view that aquifers are similar to mineral resources and are governed by this resolution. Accordingly, they insisted on the inclusion of this sovereignty article. It is noted that second sentence of Article 3 states that an aquifer state shall exercise its sovereignty in accordance with international law and the present Draft Articles. I believe that the current draft Article 3 represents an appropriately balanced text.

Article 4: equitable and reasonable utilization

The principle of equitable and reasonable utilization of shared resources has now been firmly established in international law. Factors relevant to such equitable and reasonable utilization of aquifers are listed in Article 5. Although the term is the same as that used in the International Waterourses Convention, the principle of reasonable utilization is quite different here. It covers non-renewable resources as well as renewable ones, while the principle of sustainable utilization applies only to renewable resources. International law has developed the precise legal concept of sustainability in relation to marine living resources. The principle of sustainable utilization is contained in almost all the fisheries conventions. This principle is clearly defined as "to take measures, on the best scientific evidence, to maintain or restore populations of harvested species at levels which can produce the maximum sustainable yield" in Article 119, sub-paragraph 1 (a) of the UN Convention on the Law of the Sea (UNCLOS). Science tells us that such a level of maximum sustainable yield is somewhat below the maximum population of a particular fish stock that nature can maintain. This principle can be applied to other renewable resources.

The 1997 International Watercourses Convention applied this principle and defined it as "optimal and sustainable" utilization in its Article 5, sub-paragraph 1. This obliged watercourse states to limit the amount of water used to that of river recharge and to keep the river flowing permanently. When we considered the dynamics of aquifers, it became clear that this principle could not apply to non-recharging aquifers where any utilization depletes the resource, just as for the mineral resources such as coal. Even for recharging aquifers, the consensus in the ILC was to apply a more flexible standard that would not deprive states of the use of the accumulated resources of non-recharging aquifers while allowing similar use from recharging aquifers. Accordingly Article 4 does not refer to sustainability at all and provides that a recharging aquifer shall not be utilized at a level that would prevent continuance of its effective functioning. Meanwhile, the term "sustainability" has become a sort of catchphrase for many environmentalists. Taking into account their position, the term "sustainable development" is inserted in Article 7, relating to the general obligation to cooperate.

Article 6: obligation not to cause significant harm

Another important principle is the obligation not to cause significant harm to other states. This is the cardinal principle of international law. This principle applies not only to adverse effects to other states caused by utilization of aquifers but also to adverse effects through aquifers to other states caused by other activities as defined in subparagraph (b) of Article 1. However, the utilization of aquifers and other activities are necessary for any community. Other states, therefore, have an obligation to bear certain harm unless it exceeds the level of significant harm. The concept of "significant" is relative and cannot be defined abstractly. However, in view of the fragility of aquifers and the difficulty of removing pollutants from aquifers once affected, the threshold of "significant" is much lower than in the case of surface waters. Article 15 provides a consultation procedure for a planned activity which may affect a transboundary aquifer and thereby may have a significant adverse effect upon another state.

Articles 7–15: international cooperation

Yet another important principle is that of international cooperation. The key to the proper management of aquifers is international cooperation among aquifer states. The Draft Articles provide various measures beginning with the regular exchange of data and information, monitoring and establishment of joint management mechanisms.

Article 16: technical cooperation

The Draft Articles also regulate non-aquifer states. In particular, all states are required to promote technical cooperation with developing states in the scientific, educational, technical, legal and other fields for the protection and management of aquifers in Article 16. Here, for example, Japan, although not a transboundary aquifer state, could play a significant role.

The Draft Articles and the 1997 International Watercourses Convention

The relationship between the present Draft Articles and the 1997 International Watercourses Convention and other treaties is left for future negotiation after the present Draft Articles are finalized. It is generally understood that the 1997 International Watercourses Convention covers groundwater that is linked to an international watercourse. Its Article 2 (a) provides that: "Watercourses mean a system of surface waters and groundwaters constituting by virtue of their physical relationship a unitary whole and normally flowing into a common terminus." It is not clear what kind of relationship between surface water and groundwater is envisaged here. The present Draft Articles cover all transboundary aquifers regardless of whether they are recharged by or discharged to an international watercourse.

It must also be pointed out that according to the definition of the present Draft Articles, recharge and discharge zones are outside of aquifers. The ILC considered that all aquifers possess characteristics distinct from those of surface waters. For instance, the Nubian Sandstone Aquifer System is linked to the River Nile south of Khartoum. However the bulk of the Nubian system is a non-recharging aquifer. Accordingly, that aquifer system must be regulated by the present Draft Articles. Since the Draft Articles are generally much broader and stricter than the 1997 International Watercourses Convention, applying both instruments to a particular aquifer should normally not cause any difficulty. However, if it does, drafting a provision to regulate the relationship between the two instruments is required.

Looking forward

The UN General Assembly received the Draft Articles favourably in 2008. I am convinced that this was because it recognized that they are not only scientifically and technically sound, but also incorporate the positions of the majority of the member states of the United Nations. It adopted resolution 63/124 entitled "the Law of Transboundary Aquifers" by consensus on 11 December 2008. The resolution took note of the Draft Articles, encouraged the states concerned to make appropriate bilateral or regional arrangements for the proper management of their transboundary aquifers, taking into account the provisions of the Draft Articles, and further decided to include in the provisional agenda of its 66th session in Autumn 2011 an item entitled "the Law of Transboundary Aquifers" with a view to examining, inter alia, the question of the form that might be given to the Draft Articles. In other words, consideration would be given to whether the Draft Articles should be transformed into a convention or comparable document. States must realize the grave situation of many transboundary aquifers, increase understanding and appreciation of the Draft Articles, and act in the best way to properly manage their transboundary aquifers through international cooperation. It is gratifying to note that the states of the Guarani Aquifer system have already implemented the UN resolution by adopting a new agreement incorporating the principles embodied in the Draft Articles.

Notes

1. See http://www.igrac.net/dynamics/modules/SFIL0100/view.php?fil_Id=121.
2. See FAO (2005).
3. For the text of the Draft Articles see: http://untreaty.un.org/ilc/texts/instruments/english/draft%20 articles/8_5_2008.pdf. For the commentaries see: http://untreaty.un.org/ilc/texts/instruments/ english/commentaries/8_5_2008.pdf.

Reference

FAO, 2005. *Groundwater in international law: compilation of treaties and other legal instruments.* Legislative Study 86. Rome: FAO.

Managing buried treasure across frontiers: the international Law of Transboundary Aquifers

Gabriel E. Eckstein

In recent years, transboundary aquifers have received growing attention in numerous policy-making and negotiating circles. This development suggests an evolution in customary international law for transboundary aquifers. This article examines recent international arrangements and pronouncements related to the assessment, use, allocation and protection of transboundary groundwater resources and identifies the legal trends emerging from these instruments. The article also considers gaps and shortcomings in the emerging regulatory regime and offers recommendations for the further development of the law.

Introduction

Until quite recently, transboundary groundwater resources were treated as the neglected stepchild of international water law.[1] Transboundary aquifers were habitually ignored in projects with international implications, consistently omitted from treaties and cursorily misunderstood in much of legal discourse. To a large extent, these resources were "out of sight, out of mind", largely because few realized that the resource they pumped was shared with another county. This neglect now appears to be on the wane.

Over the past few decades, transboundary aquifers have received increasing attention at all levels of civil society in the context of policy and law-making initiatives, academic exercises and a number of significant negotiations. A highly detailed management and allocation scheme was implemented on the Genevese Aquifer situated along the French–Swiss border (Convention Genevois 2008), while more rudimentary consultative and data-sharing agreements were applied to the Nubian Sandstone and Northwestern Sahara aquifers in North Africa (NSAS 2002, SASS 2002); a comprehensive cooperative arrangement was drafted for the Iullemeden Aquifer System in West Africa (Iullemeden MoU 2009), while a less-specific agreement was formulated for the nations overlying the Guarani Aquifer in South America (Guarani Acuerdo 2010); and a model transboundary-aquifer agreement was developed by a multidisciplinary group of experts meeting in Bellagio, Italy, for the Mexico–United States border (Hayton and Utton 1989). Furthermore, transboundary groundwater resources have featured prominently in the 1992

United Nations Economic Commission for Europe Convention on the Protection and Use of Transboundary Watercourses and International Lakes (UN/ECE 1992), the 1997 UN Convention on the Non-Navigational Uses of International Watercourses (UN 1997), the 2000 Revised Protocol on Shared Watercourses in the Southern African Development Community (SADC 2000) and the International Law Association's 2004 Berlin Rules on Water Resources Law (ILA 2004). In addition, unofficial arrangements have been forged by sub-national political entities for the Hueco Bolson Aquifer underlying the cities of El Paso and Juárez on the Mexico–United States border (Juárez–El Paso MoU 1999), and for the Abbotsford–Sumas Aquifer between the US State of Washington and Canadian province of British Columbia (Abbotsford–Sumas MoA 1996). And in late 2008, the UN International Law Commission (UNILC) formulated 19 draft articles on the law of transboundary aquifers (Draft Articles) thereby focusing global attention on the legal status of aquifers that traverse international political boundaries (UNGA Resolution 2008). While this catalogue is not comprehensive, the breadth of this list indicates that concerns about aquifers shared by multiple nations are no longer secondary to those of surface-water resources. Transboundary aquifers have come into their own and are now legitimate topics of international law, policy and relations.

The escalating interest in transboundary groundwater resources is largely a reflection of the growing importance that aquifers, especially those traversing international boundaries, are having in meeting nations' development needs and objectives. This is particularly evident in the Middle East where the use and allocation of transboundary aquifers, such as the Qa Disi Aquifer shared by Jordanians and Saudi Arabians and the Mountain Aquifer shared by Israelis and Palestinians, continue to be a source of friction among overlying riparians. The growing interest in shared groundwater resources, however, is also a response to the realization that few concrete rules exist in international law to govern relations over these buried treasures. As nations around the world begin to extract (or intensify their withdrawals of) groundwater from aquifers traversing international political boundaries, they are raising important questions related to the rights and obligations that riparian states can claim to these shared resources.

State practice and experience pertaining to the utilization of transboundary aquifers are slowly becoming more prevalent, and nations and scholars are now giving more thought to the formulation of rules and regulations for the management of these critical resources. While the law of transboundary aquifers is in an early stage of development, it is already possible to identify a number of trends, based on the practices of states, which are relevant to this nascent body of law. In an effort to identify and characterize these trends, this study reviews some of the activities and procedures states have pursued in the coordination, management and allocation of transboundary groundwater resources as evinced by formal and informal arrangements that they have entered into in relation to those resources. It also considers the work-product of international organizations relevant to the topic.

The study begins by addressing the importance of transboundary aquifers as a source of fresh water for people and the environment. It then examines a number of specific legal arrangements developed by nations for the assessment, use, allocation and protection of transboundary groundwater resources. By studying these regimes and the practices of states, this article attempts to identify trends that implicate the emergence of generally accepted international legal norms applicable to transboundary aquifers. This study concludes that while the law of transboundary aquifers is in an early stage of development, there is, nonetheless, a growing body of experience and practice indicating the emergence of accepted legal standards. Lastly, the article considers the gaps and shortcomings in the emerging international regulatory system and offers recommendations for the further development of the law.

The importance of transboundary aquifers

In places like North Africa, the Middle East and the Mexico–United States border, transboundary aquifers often serve as the primary or sole source of fresh water for human and environmental sustenance. Libya, for example, which has no meaningful surface-water resources, obtains the majority of its fresh water – some 6.5 million cubic metres of water daily – from the Nubian Sandstone Aquifer, a vast underground reservoir that also underlies sections of Libya's neighbours (Chad, Egypt and Sudan) (Watkins 2006). Similarly, Palestinians in the West Bank and Gaza obtain the great majority of their water from aquifers shared with Israel: the Mountain Aquifer underlying the West Bank and eastern Israel and the Coastal Aquifer underlying Gaza's and Israel's Mediterranean coast (World Bank 2009). In addition, transboundary aquifers serve as the sole source of fresh water for many of the communities along the Mexico–United States border, including the Mexican cities of Puerto Palomas, Naco, Nogales, Sonoyta and Tecate, and their respective American sister cities of Columbus, Bisbee, Nogales, Lukeville and Tecate (Eckstein 2011).

Although the exact number of aquifers traversing international political boundaries around the world is still unknown, transboundary aquifers underlie the territory of nearly every non-island nation (Puri and Aureli 2009). An ongoing study has, to date, catalogued 273 transboundary aquifers globally (UNESCOPRESS 2008), including more than 155 on the European continent, 38 in Africa, 68 in the Americas and 12 in Asia (Stephan 2009). In comparison, there are 263 watercourses crossing national boundaries (UNEP 2002). Additionally, while the extent to which states and peoples around the world rely on groundwater resources traversing sovereign borders has yet to be precisely quantified, groundwater today is the most extracted natural resource in the world. Groundwater provides more than half of humanity's freshwater for everyday uses such as drinking, cooking and hygiene, as well as 20% of irrigated agriculture (UNESCO and World Water Assessment Programme 2003). In Europe, between 60% and 99% of drinking water comes from groundwater (Almássy and Busás 1999); in the United States, that figure is between 50% and 97% (Burchi 1999). Accordingly, it is now quite evident that groundwater resources generally, and transboundary aquifers specifically, are a crucially important source of fresh water globally, and that the sound management and regulation of these shared resources are critical for ensuring the sustainability of human communities and the environment.

Examples of regulatory mechanisms for transboundary aquifers

Historically, groundwater resources were omitted from or neglected under international law and cursorily misunderstood within the legal community (Eckstein 2003). This, in turn, resulted in a blatant disregard for subsurface waters in projects with transboundary implications. Over the past few decades, though, largely in response to growing populations and expanding economies, interest in groundwaters traversing international borders has increased considerably, especially as nations intensified withdrawals from known aquifers and expanded their search for new sources of fresh water. As a result, there is now a growing international concern with the use, allocation, management and protection of transboundary aquifers. In a number of cases, this interest has spawned coordinated efforts and mechanisms, of which the most notable were identified in the Introduction. Given space limitations, only four of these efforts are examined in more detail.

The best known, and still the only treaty managing and specifically allocating the waters of a transboundary aquifer, is the Convention relative a la protection, a l'utilisation, a la realimentation et au suivi de la Nappe Souterraine Franco-Suisse du Genevois (Genevese

Agreement), commenced in 1978 and revised in 2008 (Convention Genevois 2008). This singular arrangement addresses groundwater quality, quantity, abstraction and recharge largely through the creation of a joint Genevese Aquifer Management Commission. While the Commission has only consultative status, its recommendations and technical opinions carry considerable weight in the modification of existing and development of new water-extraction works. The regime also recognizes the Swiss artificial-recharge obligations created in the original 1978 agreement, allocates expenses between the countries for the Swiss recharge efforts, and places strict withdrawals limits on the French side (see Preamble, Arts. 2.3, 8, 11–14, and Annex to the Convention on the Inventory of the recharge equipment and existing extraction works). The Genevese Agreement is particularly significant because it strikes a balance between state sovereignty and state responsibility in its management scheme, which is based almost exclusively on principles of transparency, good faith and cooperation. Moreover, the agreement is unique in proffering a purely technical mechanism that is devoid of any provisions directly related to nations' sovereign rights to the underlying aquifer or its waters.

Another fascinating arrangement, albeit one still undergoing development, is the 2009 Memorandum of Understanding Relating to the Setting up of a Consultative Mechanism for the Management of the Iullemeden Aquifer System (Iullemeden MoU) entered into by Mali, Niger and the Republic of Nigeria (Iullemeden MoU 2009). While not a binding instrument, the details and language used in the Iullemeden MoU reflect an intention by the parties to comply with the terms of the resulting agreement once it comes into force. As noted by its title, the focus of this arrangement is the creation of a Consultative Mechanism tasked with promoting cooperation over the management of the Iullemeden. More specifically, the Mechanism is responsible for, inter alia, formulating opinions related to water-management and utilization operations, coordinating Iullemeden-related activities and harmonizing procedures and policies, and developing action plans for implementing its recommendations (see Art. 5). In contrast to the purely consultative status of the Genevese Aquifer Management Commission, the Consultative Mechanism would have legal personality and authority to contract, acquire and dispose of property, seek and obtain loans, gifts and technical assistance, and be a party in legal proceedings (see Art. 6). In addition, and in stark contrast to the Genevese Agreement, the Iullemeden Memorandum explicitly relies on the well-known international water-law and environmental-law principles of equitable and reasonable utilization, no harm, exchange of data and information, prior notification, protection of the environment, public participation, precautionary approach, and polluter and user pays (see Arts. 13–17, 19–22 and 24). Nonetheless, like the Genevese Agreement, the Iullemeden Memorandum focuses more on balancing state sovereignty and state responsibility, yet avoids any explicit mention of sovereignty in its formulation.

The memorandum of understanding (Juárez-El Paso MoU) entered into by the municipal water utilities of the City of Juárez in Mexico and the City of El Paso in Texas, United States (Juárez–El Paso MoU 1999) is an especially unique transboundary aquifer arrangement. This mechanism is distinctive in that it was entered into by sub-national political entities without the oversight of the respective federal governments. While legally unofficial and unenforceable, the purpose of the arrangement is to encourage cooperation over the management and exploitation of the Hueco Bolson Aquifer, which as a result of population growth and development in the overlying region has been overexploited, raising concerns for the aquifer's viability as an ongoing source of fresh water for the area (Eckstein and Hardberger 2008). The Juárez-El Paso MoU "seeks to identify the mechanisms between the parties to increase communications, cooperation, and implementation of transboundary projects of common interest". Moreover, in its "general objectives", the arrangement alludes to data and information sharing related to transboundary natural resources, and cooperation in the management, use and protection of natural resources that traverse an international boundary (see final paragraph of Recitals). Furthermore, the MoU obligates the sister cities to develop and coordinate a compatible plan "to secure water supplies and extend the life of the Hueco Bolson" (see para. 3[a]). Given its parochial origin and perspective, it is understandable that the Juárez–El Paso MoU makes no reference to principles of international law or notions of sovereignty.

Possibly the most profound effort to develop an international regulatory system for transboundary aquifers is that undertaken by the UN International Law Commission (UNILC). In late 2008, following six years of intense work, the UN General Assembly (UNGA) adopted a resolution containing 19 draft articles on transboundary aquifers that were prepared by the UNILC (UNGA Resolution 2008). Modelled largely on the 1997 Watercourses Convention, the chief substantive state obligations are the well-respected rules of equitable and reasonable utilization and no significant harm. Those principles, however, are tailored to the unique qualities that differentiate surface waters from groundwater resources. For example, the list of factors for assessing what constitutes equitable and reasonable utilization includes "the natural characteristics of the aquifer or aquifer system", "the contribution to the formation and recharge of the aquifer or aquifer system", and "the role of the aquifer or aquifer system in the related ecosystem" (see Arts. 5[1][c], [d] and [i]). Likewise, the no significant harm rule obligates aquifer states not to cause significant harm through "activities other than utilization of a transboundary aquifer . . . that have, or are likely to have, an impact upon that transboundary aquifer" (see Art. 6). This latter modification specifically relates to the distinct likelihood that non-aquifer utilization activities undertaken above or around aquifers could detrimentally affect those aquifers, such as industrial and agricultural operations in the recharge zone, mining activities in the aquifer matrix, and construction, forestry and other activities that might affect the normal recharge process (Eckstein 2007). Other notions found in the Resolution include obligations to regularly exchange data and information, safeguard ecosystems, protect recharge and discharge zones, prevent pollution, monitor the aquifer and prior notification of planned activities (see Arts. 8, 10, 11, 12, 13 and 15). The future form (that is, convention or guidance document) of the 19 articles has been scheduled for consideration by the UN General Assembly at its 66th session in autumn of 2011 (UNGA Resolution 2008).

Developing international law for transboundary groundwater resources

The international law for managing and allocating transboundary groundwater resources is still in a nascent state. Nevertheless, the increasing number of agreements and related activities, as well as the growing international interest in the subject, suggests the evolution of customary international law in this area. Customary international law refers to international law that is based on accepted practice rather than codified rules. It emerges from the broad and consistent conduct of states that is undertaken by a belief that such behaviour is both legally appropriate and mandated (Brownlie 1998). While the extent of the practice relating to the management of a transboundary aquifer is still somewhat limited, a review of the arrangements discussed above nevertheless indicates a number of emerging trends in customary law applicable to transboundary aquifers. In particular, three procedural and two substantive norms can be identified.

Possibly the most palpable customary obligation to emerge from the evolving state practice is the procedural duty to regularly exchange data and information about a transboundary aquifer. Appearing in all of the arrangements discussed in this study, the duty is fundamental to the cooperation over and sound management and protection of transboundary aquifers. Without the sharing of such information, aquifer states' activities will be hampered by an inability to fully project and mitigate any deleterious consequences that might result from the utilization of a particular transboundary aquifer (Eckstein 2007). Accordingly, the duty requires aquifer states to share collected information on the aquifer and its functioning on a continuing basis. While the precise types of data and information that must be shared is not always spelled out, it is obvious that they should relate to the character, use and functioning of the aquifer. Draft Article 8 of the UNGA Resolution on the law of transboundary aquifers provides that it should include material of a "geological, hydrogeological, hydrological, meteorological and ecological nature and related to the hydrochemistry of the aquifers or aquifer systems, as well as related forecasts" (UNGA Resolution 2008).

A corollary procedural obligation to the duty to regularly exchange data and information is the requirement to generate supplemental data and information on an ongoing basis through monitoring and related activities. Found in a majority of the arrangements considered in this study, this obligation is indispensible to fulfilling the duty to exchange data and information. The obligation also acknowledges the need to maintain vigilance in managing a transboundary aquifer and the need to continuously check on activities related to the aquifer's utilization and the possible impact they may have on people and the environment. Under the UNGA Resolution on the Law of Transboundary Aquifers, aquifer atates "shall, wherever possible, carry out these monitoring activities jointly with other aquifer States concerned and, where appropriate, in collaboration with competent international organizations" (see Art. 13). Implementing this approach in some detail, the Genevese Agreement mandates, inter alia, that all quantitative and qualitative monitoring activities "shall be performed jointly by Swiss and French authorities on their respective territories", and water pollution analyses "shall be made at regular intervals" (see Arts. 10 and 16).

The obligation to monitor and continuously generate additional data accords with the comparable duties imposed on riparians of transboundary surfacewaters. In his separate opinion in the *Case Concerning the Gabčikovo-Nagymaros Project* before the International Court of Justice (ICJ), Judge Christopher Weeramantry argued for the emergence of a principle of continuing environmental-impact assessment. In that opinion, Judge Weeramantry opined that: "As long as a project of some magnitude is in operation, [an environmental-impact assessment] must continue, for every such project can have unexpected consequences; and considerations of prudence would point to the need for continuous monitoring" (ICJ 1997, p. 111). More recently, in the *Case Concerning the Pulp Mills on the River Uruguay*, the ICJ asserted that: "Once operations have started and, where necessary, throughout the life of the project, continuous monitoring of its effects on the environment shall be undertaken" (ICJ 2010, p. 205). While both cases applied this recurring obligation in the context of the use and development of a transboundary watercourse, the logic applied is equally and undeniably applicable to transboundary groundwater resources.

Another related procedural obligation found in a majority of the above-noted instruments is the duty of prior notification of planned activities. In essence, where a planned project has the potential to adversely affect either the territory of another aquifer state or the transboundary aquifer itself, the acting state must notify other aquifer states of its plans. This obligation is designed to allow potentially affected states to evaluate the possible consequences and to seek an understanding or compromise with the acting state (Eckstein 2007). While the precise procedures required under this principle vary among the instruments, the basic notions of advance notification are well accepted in international water law. Under the UNGA Resolution on the Law of Transboundary Aquifers, aquifer states would be obligated to provide "timely" notification "accompanied by available technical data and information . . . to enable the notified State to evaluate the possible effects of the planned activities" (see Art. 15). In contrast, the Iullemeden MoU proffers a more rigorous process of notification that includes: provisions relating to a time period during which the acting state must refrain from proceeding with the planned activity while the Consultative Mechanisms and the informed state review the information; authorization to proceed with the planned activity in the absence of a response from the Consultative Mechanism; consultation and negotiations in good faith where the acting and notified states disagree about the expected consequences of the planned activity; measures in the absence of prior notification; and an emergency exception to these obligations (see Arts. 24–28).

Among the substantive obligations found in the various arrangements discussed in this article, two well-known rules appear in a majority of instruments; those of equitable and reasonable utilization and of no significant harm. Recognized broadly as the cornerstones of transboundary surface water law, these two principles now appear to have been extended to transboundary aquifers through various academic and legalistic pronouncements. According to the first doctrine, aquifer states must ensure that their utilization of transboundary aquifers is both equitable, in terms of the benefits derived from the use of the aquifer, and reasonable with regard to the use itself. A non-exhaustive list of factors is typically provided in these instruments to aid in determining whether a particular use confirms to these criteria. Similarly, aquifer states are bound to ensure that their activities related to shared groundwater resources do not result in significant harm to other aquifer states. Like their application in the context of surface waters, none of the transboundary aquifer arrangements elaborate on the implementation of these principles. Nonetheless, they represent the nascent state of the law of transboundary aquifers.

Although the instruments discussed in this article present other rules and procedures for transboundary aquifers, the lack of their consistent appearance across the various instruments and arrangements suggest that these additional concepts do not yet represent a trend in the development of customary international law for transboundary aquifers. Nonetheless, the presence of these additional rules and procedures – such as obligations to safeguard aquifer-dependent ecosystems, protect the recharge and discharge zones of aquifers and prevent aquifer pollution – should not be discounted. Rather, their use and effectiveness as regulatory and management mechanisms should be reviewed and assessed periodically, as should their applicability for other transboundary-aquifer agreements.

Considerations for the continued development of international law for transboundary groundwater resources

While surface and groundwater resources share numerous similarities, groundwater possesses a number of unique characteristics that must be considered carefully when contemplating regulatory tools for managing these resources. For example, groundwater is typically more vulnerable than surface water to agricultural, industrial and municipal pollution as well as other forms of contamination. This is so because water in aquifers generally flows at much slower rates than in rivers and lakes, typically measured in distances of centimetres or metres per day (Hamblin and Christiansen 2001). This slower flow rate greatly diminishes the natural filtering capacities of aquifers and, thereby, their ability to reclaim and cleanse themselves. In addition, because of the geographic extent of most aquifers and the difficulties associated with monitoring and working with underground formations, the artificial reclamation of a polluted aquifer can be prohibitively complex and expensive. The result is that once contaminated, an aquifer may be rendered unusable for years, decades or longer (Eckstein 2007).

Accordingly, these unique aspects must be taken into account in order to formulate appropriate regulatory mechanisms for the sound management of transboundary aquifers. For example, special attention must be paid to the "functioning" of an aquifer, which refers to how particular aquifers work or behave as aquifers. Aquifers typically store and transport water, dilute wastes and other contaminants, provide a habitat for aquatic biota and serve as a source of fresh water and nutrients to aquifer-dependent ecosystems. Some aquifers even provide geothermal heat. Each of these characteristics comprises a function of the aquifer that is dependent on the particular aquifer's hydrostatic pressure, hydraulic conductivity, and mineralogical, biological and chemical attributes, all of which may be interdependent with each other to the extent that the aquifer's sustained operation as a dynamic hydrogeologic system depends on the continuation of a particular function or series of functions (Heath 2004). If any component related to an aquifer's natural behaviour is impaired or destroyed, it could result in serious consequences for the viability and integrity of the aquifer as a whole, as well as to communities and ecosystems dependent on that aquifer. Accordingly, regulatory mechanisms must be developed to ensure the identification, maintenance and protection of the functioning of transboundary aquifers that incorporate functional characteristics on both sides of a border, tasks that can best be achieved in a coordinated and cooperative process.

Highly related to the functioning of aquifers, recharge and discharge zones of aquifers also require special attention since, with the exception of some deep, non-recharging, aquifers, these zones are integral components of aquifers. Recharge and discharge zones regulate the flow and quality of water moving into and out of aquifers and, thereby, the functioning of aquifers themselves. Hence, the recharge and discharge process, as well as the geographical area in which they operate, must be maintained and protected. In the case of recharge zones, this consists of ensuring both the quantity and quality of water flowing through the recharge zone and entering the aquifer. Thus, recharge-zone protection might include limitations on industrial and municipal development projects in the recharge area that could diminish the amount or affect the quality of water percolating through the zone into the aquifer. It also might include restrictions on industrial and agricultural activities that might pollute the recharge area and thereby pollute the water flowing into the aquifer. In the case of discharge zones, protective action could include restrictions on construction and other activities that could detrimentally affect the discharge process, water flow within the aquifer, the location of the water table or the aquifer's natural cleansing abilities. In both zones, it could also include constraints on mining activities that remove or modify strata within the recharge and discharge areas. Moreover, limitations in the two zones should be pursued regardless of whether or not the targeted activities are related to the use of the aquifer itself.

Finally, a conceptual notion that is inherently tied to the transboundary characteristic of these aquifers pertains to the perception, held by some nations, that they have unqualified or near-absolute sovereignty over the portion of an aquifer that lies within their jurisdiction. The belief, which harkens back to the long-discredited Harmon Doctrine, suggested that states are free to exploit resources within their jurisdiction without regard to the extraterritorial effects of such action (McCaffrey 2007). While a number of the instruments reviewed here appear to subscribe to this notion, the suggestion that a liquid resource – which is in a near-constant state of motion moving within the hydrologic cycle in surface waters, subsurface strata and the atmosphere, transforming between gaseous, liquid and solid states, and traversing local and international political borders – can be subject to a state's sovereignty defies logic. Moreover, the idea is contrary to the basic tenets of international water law, including those of equitable and reasonable utilization and of no significant harm, which clearly espouse a more limited conception of sovereign rights over transboundary freshwater resources (McCaffrey 2011).

While the Draft Articles on Transboundary Aquifers adopts the notion of sovereignty in Draft Article 3, it also includes a provision in the same article that could temper states' claims and obligate them to exercise their sovereign rights "in accordance with international law and the present articles" (UNGA 2008). In essence, states that agree to the terms of the Draft Articles would relinquish some measure of sovereignty to the extent that they give up their right to act contrary to other provisions and obligations contained in the Draft Articles. This is because the principles contained in the Draft Articles – such as equitable and reasonable utilization, no significant harm, exchange of data, monitoring and others – clearly place considerable restrictions and obligations on what aquifer states can do with regard to the utilization of a transboundary aquifer (Eckstein 2007). While this saving grace may be unsatisfactory for some (McCaffrey 2011), others suggest that it reflects an unavoidable compromise between nations seeking to preserve sovereignty over resources found within their territory, and those who argued for a more collaborative approach to resources that traverse international boundaries (Yamada 2011).

Conclusion

Until quite recently, concerns over aquifers traversing international boundaries were rarely raised in international fora. With the advent of drilling technology and the growing global need for additional sources of fresh water, transboundary aquifers are now the focus of considerable international attention, including scrutiny by the UN General Assembly. As such, they are also becoming the focus of new treaties and other arrangements and, thereby, the development of international standards. While this evolutionary legal process is still in its infancy, a number of trends can be identified from the practices of states (as reflected in instruments formulated to manage specific transboundary aquifers) that indicate the emergence of customary international norms. One of the purposes of this study was to identify these nascent trends.

As the number of such arrangements grows, as is certain to happen, many of these trends will become more evident and will lead to the development of a regime for the management of transboundary groundwater resources. In the meantime, numerous gaps and shortcomings exist in the absence of such a regime that must be considered, as states formulate new agreements for other transboundary aquifers. Hence, the second purpose of this article was to develop recommendations for the further development of the law.

The recommendations offered here, however, do not cover the gamut of issues and aquifer characteristics that require attention within an international regulatory and legal context. Others concerns and topics that should be considered include, inter alia: the threshold of harm necessary to trigger a violation of the no significant harm standard as it applies to transboundary aquifers; the aspects of an aquifer, in terms of functions and geographic scope, that should be covered by a regulatory or cooperative transboundary regime; harmonization of metadata and methodologies among aquifer riparians for generating information about shared aquifers; and the rules that should govern the exploitation of non-recharging aquifers. Nevertheless, this article provides a starting point from which to further the conversation about the characteristics of transboundary aquifers that must be considered, as well as the regulatory and legal mechanisms that might develop.

Note

1. This article expands on a prior paper that appears in UNESCO (2010).

References

Abbotsford–Sumas MoA, 1996. *Memorandum of Agreement related to referral of water right applications related to the transboundary Abbotsford–Sumas Aquifer between the State of Washington as represented by the Department of Ecology and the Province of British Columbia as represented by the Minister of Environment, Lands and Parks, 10 October 1996* [online]. Available from: http://internationalwaterlaw.org/documents/regionaldocs/Local-GW-Agreements/1996-BC-WA-Water-Right-Referral-Agreement.pdf [Accessed 16 June 2011].

Almássy, E. and Busás, Zs., 1999. *UN/ECE task force on monitoring & assessment guidelines on transboundary ground water monitoring. Volume 1: inventory of transboundary ground waters*. Lelystad, the Netherlands: UN Economic Commission for Europe.

Brownlie, I., 1998. *Principles of public international Law.* 5th ed. New York: Oxford University Press.

Burchi, S., 1999. National regulation for groundwater: options, issues and best practices. *In*: S.M.A. Salman, ed. *Groundwater: legal and policy perspectives. Proceedings of a World Bank Seminar*. Washington, DC: The World Bank, 55–67.

Convention Genevois, 2008. *Convention relative a la protection, a l'utilisation, a la realimentation et au suivi de la Nappe Souterraine Franco-Suisse du Genevois* [online]. Geneva, 18 December 2007; in force on 1 January 2008. Available from: http://internationalwaterlaw.org/documents/regionaldocs/2008Franko-Swiss-Aquifer.pdf [Accessed 16 June 2011].

Eckstein, G., 2003. A hydrogeological approach to transboundary ground water resources and international law. *American University International Law Review*, 19 (2), 201–258.

Eckstein, G., 2007. Commentary on the UN International Law Commission's Draft Articles on the Law of Transboundary Aquifers. *Colorado Journal of International Environmental Law & Policy*, 18 (3), 537–610.

Eckstein, G., 2011. Buried treasure or buried hope? The status of Mexico–U.S. transboundary aquifers under international law. *International Community Law Review*, 13 (3), *forthcoming*.

Eckstein, G. and Hardberger, A., 2008. State practice in the management and allocation of transboundary groundwater resources in North America. *Yearbook of International Environmental Law*, 18, 96–125.

Guarani Acuerdo, 2010. *Acuerdo sobre el Acuífero Guarani* [online]. San Juan, Argentina, 2 August. Available from: http://internationalwaterlaw.org/documents/regionaldocs/Guarani_Aquifer_Agreement-Spanish.pdf [Accessed 16 June 2011].

Hamblin, W.K. and Christiansen, E.H., 2001. *Earth's dynamic systems*. Upper Saddle River, NJ: Prentice Hall.

Hayton, R. and Utton, A., 1989. Transboundary groundwaters: the Bellagio Draft Treaty. *Natural Resources Journal*, 29, 663–722.

Heath, R.C., 2004. *Basic ground-water hydrology*. Water supply paper 2220. Reston, VA: US Geological Survey. Available from: http://pubs.er.usgs.gov/djvu/WSP/wsp_2220.pdf [Accessed 18 July 2011].

ICJ, 1997. *Case concerning the Gabčikovo-Nagymaros Project. Separate opinion of Judge Weeramantry. Judgment of 25 September 1997* [online]. Available from: http://www.icj-cij.org/docket/files/92/7383.pdf [Accessed 16 June 2011].

ICJ, 2010. *Case concerning the pulp mills on the River Uruguay. Judgment of 20 April 2010* [online]. Available from: http://www.icj-cij.org/docket/files/135/15877.pdf [Accessed 16 June 2011].

ILA, 2004. *Berlin Conference on water resources law* [online]. Available from: http://internationalwaterlaw.org/documents/intldocs/ILA_Berlin_Rules-2004.pdf [Accessed 16 June 2011].

Iullemeden MoU, 2009. Memorandum of understanding relating to the setting up of a consultative mechanism for the management of the Iullemeden aquifer system. Bamako, 20 June 2009.

Juárez-El Paso MoU, 1999. *Memorandum of Understanding between City of Juárez, Mexico Utilities and the El Paso Water Utilities Public Services Board of the City of El Paso, Texas, 6 December 1999 [online]*. Available from: http://internationalwaterlaw.org/documents/regionaldocs/Local-GW-Agreements/El_Paso-Juarez_MoU.pdf [Accessed 16 June 2011].

McCaffrey, S., 2007. *The law of international watercourses*. 2nd ed. New York: Oxford University Press.

McCaffrey, S., 2011. The International Law Commission's flawed Draft Articles on the Law of Transboundary Aquifers: the way forwad. *Water International*, 36(5), 566–572.

NSAS, 2002. *Programme for the development of a regional strategy for the utilisation of the Nubian Sandstone Aquifer System (NSAS). Terms of reference for the monitoring and exchange of groundwater information of the Nubian Sandstone Aquifer System, 2002. Tripoli, 5 October 2000 [online]*. Available from: http://www.fao.org/docrep/008/y5739e/y5739e05.htm [Accessed 16 June 2011].

Puri, S. and Aureli, A., 2009. *Atlas of transboundary aquifers*. Paris: UN Educational, Scientific and Cultural Organization (UNESCO).

SADC, 2000. *Revised protocol on shared watercourses in the Southern African Development Community. Windhoek, 7 August 2000* [online]. Available from: http://internationalwaterlaw.org/documents/regionaldocs/Revised-SADC-SharedWatercourse-Protocol-2000.pdf [Accessed 16 June 2011].

SASS, 2002. Establishment of a consultation mechanism for the Northwestern Sahara Aquifer System (SASS) [online]. Rome, 19–20 December; endorsed 6 January 2003 (Algeria), 15 February 2003 (Tunisia), 23 February 2003 (Libya). Available from: http://www.fao.org/docrep/008/y5739e/y5739e05.htm#bm05.2.1 [Accessed 16 June 2011].

Stephan, R.M., ed. 2009. *Transboundary aquifers. Managing a vital resource. The UNILC Draft Articles on the Law of Transboundary Aquifers*. Paris: UNESCO.

UN, 1997. *Convention on the Law of Non-Navigational Uses of International Watercourses*. [online] G.A. Res. 51/229, UN GAOR, 51st Sess., UN Doc. A/RES/51/229 (1997). Available from: http://internationalwaterlaw.org/documents/intldocs/watercourse_conv.html [Accessed 16 June 2011].

UN/ECE, 1992. *Convention on the Protection and Use of Transboundary Watercourses and International Lakes. Helsinki, 17 March*. [online] Available from: http://www.unece.org/env/water/pdf/watercon.pdf [Accessed 16 June 2011].

UNESCO and World Water Assessment Programme, 2003. *Water for people, water for life. The United Nations world water development report*. UN World Water Assessment Programme. New York, Oxford: Berghahn.

UNESCO, 2010. Pre-proceedings of the conference: transboundary aquifers. Challenges and new directions, 6–8 December 2010, Paris, France. CD rom.

UNESCOPRESS, 2008. *UNESCO publishes first world map of underground transboundary aquifers*. Press Release No. 2008-108, 22 October 2008. Available from: http://portal.unesco.org/en/ev.php-URL_ID%3D43767&URL_DO%3DDO_TOPIC&URL_SECTION%3D201.html [Accessed 18 July 2011].

UNEP, 2002. *Atlas of international freshwater agreements*. UN Environmental Programme.

UNGA, 2008. *UN General Assembly Resolution on the Law of Transboundary Aquifers*. A/RES/63/124. Available from: http://internationalwaterlaw.org/documents/intldocs/UNGA_Resolution_on_Law_of_Transboundary_Aquifers.pdf [Accessed 16 June 2011].

Watkins, J., 2006. Libya's thirst for "fossil water"[online]. *BBC News*, 18 March. Available from: http://news.bbc.co.uk/2/hi/science/nature/4814988.stm [Accessed 14 June 2011].

World Bank, 2009. *Assessment of restrictions on Palestinian water sector development*. Report No. 47657-GZ. Available at: http://siteresources.worldbank.org/INTWESTBANKGAZA/Resources/WaterRestrictionsReport18Apr2009.pdf [Accessed 18 July 2011].

Yamada, C., 2011. Codification of the Law of Transboundary Aquifers (Groundwaters) by the United Nations. *Water International*, 36(5), 557–565.

The customary law applicable to internationally shared groundwater

Joseph W. Dellapenna

In 2008, the United Nations' International Law Commission (ILC) completed work on its Draft Articles on the Law of Transboundary Aquifers (ILC 2008). The United Nations General Assembly adopted a resolution later that year "taking note" of the articles and "encouraging" member states to consider the articles in making arrangements for their transboundary aquifers. Four years earlier, the International Law Association completed its own codification of this same body of law. This article explores the nature of customary international law, its application to internationally shared groundwater and the relative merits of these two codifications of the customary international law of groundwater.

Introduction

Groundwater makes up about 97% of the world's unfrozen freshwater (Dellapenna 2010), yet until recently, nations have made few bilateral or multilateral agreements regarding internationally shared groundwater. This is in sharp contrast to the numerous agreements regarding the sharing of surface water sources. Such a pattern is not surprising. Before the spread of advanced pumps after the Second World War, groundwater was strictly a local resource that could not be used in large enough volumes to affect users at any considerable distance away from the water withdrawal (Dellapenna 2010). Newer technologies and exponential growth in demand for water combined to make groundwater a critically important resource and the focus of disputes between nations. Groundwater has belatedly become the focus of a small but growing number of regional or bilateral international agreements (Burchi and Mechlem 2005, Zeitoun 2007). There are no generally applicable agreements for internationally shared groundwater. Some readers will think at this point of the United Nations (UN) Convention on the Law of Non-Navigational Uses of International Watercourses, which would apply to groundwater that is "tributary to" international watercourses (UN 1997, art. 2[a]). This convention has not yet entered into force and therefore does not apply to any water at this time, and will not apply, except perhaps as evidence of customary international law, to any waters for some time to come. The International Law Commission's (ILC) Draft Articles on Transboundary Aquifers (ILC 2008) have even less effect at this point.

This makes customary international law of even greater significance than it is for surface waters. The relevant customary law derives from the limited practice among states relative to groundwater and from international environmental law and international human-rights law (ILA 2004, introductory note). The resulting body of law applies both to internationally shared groundwater and to groundwater generally. This article, after briefly discussing the extent of existing international agreements that appear to regulate internationally shared groundwater, discusses the nature of customary international law, the patterns of agreements between states that validate the applicability of that law to internationally shared groundwater, and the recent attempts to summarise this law in an authoritative statement.

International agreements

Only rarely do international agreements relating to water mention groundwater expressly. Initially, if groundwater was mentioned at all, it was by inclusion in general water treaties, but often only by being mentioned in the definition of the waters within the scope of the agreement, without any provisions specific to groundwater issues; the focus of these agreements remained fixed on shared surface waters (Burchi and Mechlem 2005). Groundwater appeared as an afterthought even in the occasional treaty that mentions groundwater in its specific operative provisions. The gradual spread of these agreements did serve to establish that, at least in a general way, the same general legal principles apply to groundwater as apply to surface water (Eckstein and Eckstein 2003, ILA 2004, art. 36). This made sense because groundwater and surface waters are merely the same water moving through a single hydrologic cycle. Some agreements are even more vague regarding groundwater. Thus, the Mekong Agreement addresses, in certain respects, the "environment", defining that term as including "[t]he condition of water and land resources, air, flora, and fauna that exists in a particular region" (Mekong Agreement 1995, art. 2). Whether such vague agreements actually apply to groundwater as well as surface waters is at least unclear and may be doubtful.

Even growing pressure on water resources at the end of the twentieth century often did not lead to water allocation agreements specifically focused on groundwater, yet treaties did begin to address groundwater more specifically (Burchi and Mechlem 2005). Agreements referring to groundwater may be grouped into three types (Caponera and Alheritiere 1978). In some agreements, groundwater is seen as a problem impeding the development of other resources. Such agreements focus on the dewatering of aquifers. In other agreements, groundwater is seen as a resource to be developed or exploited. Such agreements focus on the allocation of groundwater withdrawals among the concerned states. Yet other agreements approach groundwater as a resource to be managed and protected. Some of these agreements focus on the prevention of harm to aquifers or to the water users who rely on the aquifers or on the management of aquifers and related resources (Kallis and Butler 2001). Such agreements did not necessarily follow any set sequence for a particular groundwater basin; nor does each groundwater basin feature all three types of agreements. Rather, states sharing a groundwater basin enter into an agreement when they see it as necessary for the particular situation, with a particular agreement sometimes including elements of several of the types of agreement.

Groundwater treaties were usually made without full knowledge of the relevant facts: the actual condition of the resource; the effect of water withdrawals from the aquifer; or the sustainability of such withdrawals. States dealt with this problem either by providing that uses shall continue in their established or historic pattern or by providing for the use of specific – often, but not always equal – quantities of groundwater by the several states without any real attempt to manage or even to assess the groundwater resource (Burchi and

Mechlem 2005). Some agreements have provided that a state would have the exclusive use of certain springs, without providing limits on the quantities to be taken. Several recent treaties simply provide for the equitable utilization of groundwater, requiring a balancing process involving the sort of factors listed in the UN Convention and the Berlin Rules – without any specific allocation and dependant on further negotiations to determine the balance of the relevant factors (Burleson 2006, pp. 388–409). Such an approach is more of an invitation to negotiation than a determination of the allocation of the shared groundwater.

Customary international law generally

To what extent, if any, does customary international law fill this void? Customary international law results from a more complex and uncertain process than formal agreements. Customary international law consists of the practices of states undertaken out of a sense that the practice is required by law (*opinio juris sive necessitatus,* or simply *opinio juris*) (Wolfke 1993, Lepard 2010). If these two elements combine, law results regardless of how long, or how briefly, the practice has continued. Customary law is binding because the participating states have expressly or implicitly consented to the rule (Wolfke 1993, pp. 160–167, Lepard 2010, pp. 97–121). Thus references to law connect a customary practice to a sense of legitimacy, and thus constitute the practice as law in a highly decentralized and institutionally undeveloped system like international law (or among subsistence farmers or nomadic tribesmen).

An analogy updated from a suggestion from some 90 years ago makes the process by which customary international law develops clearer (Cobbett 1922, pp. 1–5). Suppose, as a result of global climate disruption, people settle a newly thawed island in two villages. A meadow separates the two villages, with no road across the meadow. People initially wander at will to go from one village to the other. Gradually, most people come to follow a particular path. Perhaps it is the shortest route, or perhaps it is the easiest route, or perhaps it is the route most convenient for the heaviest walkers – walkers whose tread wears a path more decisively into the land. As a result, a definite path emerges and becomes a road. Eventually, everyone will agree that this road is the right way to travel from village to village. When people begin to object that people who follow other paths are trespassers, we have a legal and not merely a factual claim. If that claim is accepted on the island, we have a customary rule of law, even though no one can say precisely when it took hold.

Customary international law emerges in an analogous process, developing through a process of claim and counterclaim between states. When a state undertakes an action that affects other states, the other states either acquiesce in the action or take steps to oppose it, usually at first through rhetorical strategies; if the matter is important enough to an objecting state, it will escalate its opposition through various sanctions up to the possibility of military operations. Often the states involved reach a consensus, expressed through an exchange of diplomatic notes or otherwise, about what each state is entitled to do in the circumstances at hand. Beginning with the simplest rhetorical strategies and continuing through to outright war, the states involved in a controversy refer to customary international law as a justification of their claims and their practices (Byers 2000). Diplomats know very well the differences among appeals to law, appeals to morality and appeals to expedience; they often express these different propositions at appropriate points in their statements and assertions.

Customary international law works satisfactorily when there are only a few participants in a particular international process or when the law operates without major controversy – either because there is a broad consensus on what is proper or because other states are unwilling to challenge the one or few states with a strong interest in the matter (Koh 1997). Thus to determine whether customary international law exists and what is its content requires diplomats, international tribunals, lawyers and scholars to examine a wide variety of sources on state practice; finding evidence regarding the reasons for the practice is even more challenging. A widespread pattern of international agreements may demonstrate that a practice is so widely followed that it has become a rule of customary law binding even on states not party to the treaty (Wolfke 1993, pp. 68–72, Lepard 2010, pp. 191–207). Under some circumstances, even an unratified treaty can indicate customary law. General Assembly resolutions and resolutions of other international organizations are strong evidence that states consider a particular rule to be a legal obligation, leaving one to find consistent state practice, although diplomats, lawyers, scholars and even international tribunals sometimes overlook whether state practice is actually consistent with such resolutions (Wolfke 1993, pp. 79–84, 100–104, Lepard 2010, pp. 208–228). Even unilateral acts of states can demonstrate that the particular state embraces a particular customary rule of law (Rubin 1977).

Customary international law, like all customary law, is in some respects ill fitted to regulate the behaviour of states. The process of determining customary international law, even when successful, is "inelegant" (Kolb 2003). It frequently is ill defined and uncertain, and often leaves gaps and ambiguities in the law. Treaties and other international agreements only sometimes fill these gaps or clarify the ambiguities. Furthermore, turning as it does on a question of motive, examination of the primary evidence for a customary rule is often inconclusive. That, plus the lack of a neutral enforcement mechanism, makes exclusive reliance on customary international law undesirable. Yet despite the obvious difficulties in determining the precise content of customary international law, the system has been remarkably successful. Focusing exclusively on the relatively few, but highly dramatic, failures of international law can create an impression that the system is ineffective, yet no form of international life could exist without a shared set of norms that are largely self-effectuating in the conduct of that life (Rittberger and Mayer 1993). Customary international law generally serves to empower international actors by legitimating their claims, but it also constrains them by limiting the claims they can make (Norman and Trachtman 2005).

Successful areas of customary law often are codified under United Nations or other auspices, a codification possible precisely because the rules are so seldom questioned and so generally followed. There are many examples, such as the law of diplomatic immunity. The United Nations usually begins such a codification through the International Law Commission, a body created by the General Assembly in 1947 to "promote the progressive development of international law and its codification". The Commission consists of 34 jurists and diplomats representing a broad range of geographic regions, legal cultures and political ideologies. Consensus often comes after years of debate, a process that lends a high degree of credibility to the resulting codification. Upon concluding its research, the Commission reports its findings to the General Assembly, which may or may not take steps to turn the Commission's "draft articles" into a legally binding instrument. The rules assembled by the Commission often are accorded "quasi-legal effect" as rules of customary international law even before they take the form of a binding legal document. Even when a body of customary law has been codified, however, parts or even a great deal of it often survive as customary law.

Private groups also have set about to codify customary international law. One of these groups is the International Law Association, a group founded in 1873 to bring together lawyers, jurists and scholars interested in international law. It frequently presents codifications of particular fields of international law in order to provide some degree of certainty and clarity to the rules, and its efforts have often been accepted by the international community as accurately summarizing the chosen topic. One of its most successful efforts was the Helsinki Rules on the Uses of Waters of International Rivers (1966). So successful were these rules that the General Assembly instructed the International Law Commission to draft a set of articles on the subject drawing from the Helsinki Rules only four years later (UN 1970).

Customary international law and groundwater

The application of the well-established rules of customary international law to groundwater is more problematic than the application of those rules to surface waters because of the uncertain knowledge of the hydro-geologic characteristics of the resource. We know a lot about surface waters, often with accurate and ongoing measurements for a century or more. In contrast, groundwater was long seen as a mysterious substance the sources or behaviour of which were unknown. Even now we often simply do not know a great deal about particular aquifers (Dellapenna 2010). Acquiring knowledge is possible, but it is expensive and time-consuming. Because of such information problems, we often can only make tentative decisions regarding groundwater – decisions that the informal processes of customary regimes are ill adapted to revise or supplement. Customary rules or agreements to apply customary international law to groundwater do not get states very far. Proper management of groundwater, just as with surface waters, requires more than bare promises to cooperate or to take specific steps to protect and enhance the waters in question. In a fully developed management treaty, the participating states need to agree to collect and share data, to formulate joint plans and programmes for groundwater development, use and protection, to implement common groundwater management policies (including the rationing of groundwater withdrawals), and the joint training of technical personnel and the joint undertaking of environmental studies (Burleson 2006, pp. 418–423). Participating states could agree to coordinate their separate management regimes or to partial or complete joint management or they could establish a new, at least quasi-independent agency to manage the groundwater of behalf of all the participating states (Burchi and Spreij 2003, Feitelson 2003). Such fully developed treaty regimes remain rare (for example, Kemper *et al.* 2003).

The usual codification processes have been applied to water resources. The International Law Commission drafted a set of articles that the General Assembly, working through the Sixth (Legal) Committee (which includes representatives of all UN member nations), transformed into the UN Convention on the Law of Non-Navigable Uses of International Watercourses (UN 1997). This convention, which is not yet in effect, would apply only to some internationally shared groundwater, leaving out any application to groundwater that, while internationally shared, is not hydraulically connected to an international watercourse (Eckstein 2005). The convention furthermore ignores the extent to which customary international law, primarily international environmental law, applies to groundwater within a single state. In the face of these problems with the convention, both the International Law Association and the International Law Commission undertook to codify the customary international law applicable to groundwater. The International Law Association completed its project first, approving the Berlin Rules on Water Resources in 2004.

The International Law Association's decision to undertake a review and recodification of the customary international law applicable to water resources has a long and significant pedigree. The Association's Helsinki Rules (ILA 1966) had long been accepted as the definitive summary of the customary international law on internationally shared waters. The Association had also adopted a very brief set of rules in 1986 on international shared groundwater: the Seoul Rules on International Groundwaters (ILA 1986). The Seoul Rules, however, merely indicated that the rule of equitable utilization applied to groundwater without elaboration or development (ILA 1986). After eight years of work, the Association's Water Resources Law Committee produced a draft that sought to produce a comprehensive summary of all customary international law relevant to water resources – domestic, as well as internationally shared. This draft was unanimously approved by the plenary meeting of the Association in Berlin (ILA 2004).

While the Berlin Rules are an ambitious reconsideration of the entire body of international law applicable to water, including to the extent it applies to water entirely within a single state, Chapter VIII of the Berlin Rules is the first attempt at a detailed summary of the emerging customary international law specifically applicable to groundwater. Chapter VIII focuses mostly on aquifers, which are the object of management when it comes to groundwater, rather than presenting rules in terms of groundwater in the abstract (ILA 2004, arts. 36–42). While the Berlin Rules apply to all aquifers, and not just to aquifers connected to international drainage basins, the rules also indicate that the application of the general rules require attention to the special characteristics of groundwater (Berlin Rules 2004, art. 36). The Berlin Rules also recognize that aquifers, like other water sources, should be managed conjunctively and in an integrated fashion (Kallis and Butler 2001, ILA 2004, arts. 5, 6, 36;). For groundwater, the difficulty and expense of determining just what is happening underground along with the difficulty of remediating any damage that might result from poor aquifer management make application of the precautionary principle particularly compelling; therefore, the Berlin Rules express the obligation to apply the precautionary principle in strong terms for groundwater (ILA 2004, arts. 23, 38), applying this obligation to aquifers entirely within a single state because the obligation derives from international environmental law, rather than from instruments directed specifically at transboundary waters. To achieve these goals, the responsible states must acquire the necessary information (UN 1997, arts. 9, 11–19, 24[1], 28[2], 31, ILA 2004, art. 39), including by:

(1) monitoring groundwater levels, pressures and quality;
(2) developing aquifer vulnerability maps;
(3) assessing the impacts on groundwater and aquifers of industrial, agricultural, and other activities;
(4) any other measures appropriate to the circumstances of the aquifer.

The obligation to acquire the necessary information derives from, but is more specific than, the generalized obligation to gather and exchange data on transboundary waters expressed in other articles of the Berlin Rules and in the UN Convention (UN 1997, arts. 9, 11–19, 24[1], 28[2], 31, ILA 2004, arts. 56–60). Even if one concludes that the obligation to gather information is limited to transboundary contexts, often a state cannot know whether an aquifer is transboundary, with the result that a state cannot fulfil even the more limited duty unless the state develops comprehensive information for all aquifers within its jurisdiction even without proof of international connections.

In addition to the specific requirements of the Berlin Rules on groundwater, states must fulfil their general obligations to protect the aquatic environment and to respond to extreme situations (ILA 2004, arts. 22–35), including the obligation to undertake the impact assessments (ILA 2004, arts. 29–31). Three articles of the Berlin Rules' chapter on groundwater set out the specific substantive obligations of states relative to aquifers. The first of these articles addresses the application of the principle of sustainability to aquifers (ILA 2004, arts. 7, 40).

Fewer resource issues are more controversial than "groundwater mining": the view that groundwater should be viewed as an exhaustible rather than renewable resource (Chavez 2000, Albiac *et al.* 2006). Sustainability cannot altogether preclude the abstraction of water from a non-renewable aquifer, because if use is limited to the recharge rate, such waters cannot be used at all (ILA 2004, art. 40[2]). If the water is mined, the water table will decline, water pumping will become more expensive, and eventually the pumping must stop from total depletion or impaction, or because pumping becomes too expensive. In order to ensure sustainability, states should establish a maximum allowable drawdown for each aquifer, with the maximum allowable drawdown reflecting to some extent the natural and artificial recharge (ILA 2004, art. 40[1]). Treaties and other international instruments relating to internationally shared surface water and connected groundwater after 1990 stressed the concept of sustainability in just this fashion (for example, EU 2000, art. 4[1][b][ii]). Ultimately, whether to extract groundwater faster than the rate of recharge is left to the judgment of the particular states concerned (for example, EU 2000, art. 4[7], Charturvedi 2001, Fuentes 2002, Gupta 2002).

Groundwater managers must be particularly sensitive to the reality that rapidly falling water tables might not appear until some years after a serious overdraft begins, by which time it might be too late to do much about it. The precautionary principle, as well as the sustainability principle, thus requires states to be cautious in setting allowable drawdowns or in assessing other risks to the sustainability of the aquifer or of the aquatic environment related to the aquifer (ILA 2004, arts. 23, 38). Biodiversity can become an issue in setting the maximum permissible drawdown with the precautionary principle again cautioning against over-optimistic exploitation of the resource (ILA 2004, arts. 22, 25, 38). Because groundwater moves slowly and unpredictably, it is often particularly vulnerable to environmental damage and thus requires special steps to ensure the protection of aquifers (ILA 2004, art. 41). These provisions add to, without subtracting from, the general obligations to protect the aquatic environment and to respond to extreme situations (ILA 2004, arts. 22–28, 32–35). Once again, because the obligation to protect aquifers derives from international environmental law, it applies even to an aquifer entirely within a single state.

States cannot exploit more than their appropriate share of groundwater, whether from a renewable or from a non-renewable aquifer, under the principle of equitable utilization (ILA 2004, art. 42). Several international agreements now limit the drawdowns of groundwater in order to ensure that States limit their use to an equitable and reasonable share. In setting drawdown rates for transboundary aquifers, basin states are to have due regard for the obligation not to cause significant harm to another state (Article 16) and to the obligation to protect aquifers (ILA 2004, art. 41). Under Article 41(5), states are to cooperate in protecting the recharge of aquifers. Similarly, under article 41(6), the obligation to prevent significant harm applies to transboundary aquifers, having due regard to the rule of equitable utilization. States sharing an aquifer must resolve all of the questions regarding the proper management of the aquifer jointly if they are to ensure that the resource or its beneficial uses are shared equitably (ILA 2004, art. 42). Other legal principles that arguably apply to groundwater management regimes – particularly relating to public participation and related rights (ILA 2004, arts. 17–21) – find only limited support thus far in actual groundwater treaties.

The International Law Commission's Draft Articles on the Law of Transboundary Aquifers

Shortly after the General Assembly approved the UN Convention on the Law of Non-Navigable Uses of International Watercourses (UN 1997), the International Law Commission undertook to craft a set of Draft Articles on the Law of Transboundary Aquifers, completing its first reading in 2006 and its second reading in 2008 (ILC 2006, 2008, Eckstein 2007). The Commission acknowledged drawing upon the Berlin Rules and the comments of an International Law Association working group that reviewed the first reading (ILC 2008, preamble), but the Draft Articles are closer to the UN Convention's provisions on surface waters than to the Berlin Rules chapter on groundwater. Like the UN Convention rules on "international watercourses", the Draft Articles focus on the transboundary dimensions of "aquifers" and related "aquifer systems". The Draft Articles define an "aquifer" as "a permeable water-bearing underground geological formation underlain by a less permeable layer and the water contained in the saturated zone of the formation" (ILC 2008, art. 2[a]) and an "aquifer system" as "a series of two or more aquifers that are hydraulically connected" (ILC 2008, art. 2[b]). The articles focus on the rights and duties of "Aquifer States", states "in whose territory any part of a transboundary aquifer or aquifer system is situated" (ILC 2008, art. 2[d]).

The Draft Articles present 19 articles that largely track the UN Convention, particularly the principles of equitable utilization and avoidance of harm (ILC 2008, arts. 4[a], 5, 6). Other provisions track the UN Convention's duty to cooperate (ILC 2008, art. 7), the regular exchange of data and information (ILC 2008, arts. 8, 19), the protection and preservation of the environment (ILC 2008, art. 10), the prevention, reduction and control of pollution (ILC 2008, art. 12), the obligation to consult on the management of internationally shared groundwater and on planned activities likely to affect such groundwater (ILC 2008, arts. 14, 15), the obligation to cooperate regarding emergency situations (ILC 2008, art. 17), and the protection of internationally shared groundwater in time of armed conflict (ILC 2008, art. 18). Aquifer states are also required to monitor their transboundary aquifers and aquifer systems and to exchange the data resulting from the monitoring (ILC 2008, art. 13). Such a rule is necessary if the other obligations are to have real meaning. The articles, however, do not fully embrace the principle of sustainability, requiring only that aquifer states shall "aim at maximizing the long-term benefits" derived from the use of groundwater (ILC 2008, art. 4[b]), that they shall "establish . . . an overall utilization plan, taking into account present and future needs of, and alternative water sources for, the Aquifer States" (ILC 2008, art. 4[c]), and that they "shall not utilize a recharging transboundary aquifer or aquifer system at a level that would prevent continuance of its effective functioning" (ILC 2008, art. 4[d]). A number of important aspects of sustainability are omitted from these specific obligations, including the obligation to protect aquifers from pollution in order to ensure sustainability. The Draft Articles, moreover, simply do not take a clear position on groundwater mining, omitting any discussion of nonrecharging aquifers. Nor do the Draft Articles define what an aquifer's *function* is. Would exhausting it in 20 years be its function? Without some indication of what that vague word means, there is no effective limitation on exploitation and thus no real commitment to sustainability.

One of the more controversial and perplexing articles in the Draft Articles reaffirms the sovereignty of aquifer states over the portions of transboundary aquifers within their borders, while also declaring that "a State is to exercise its sovereignty in accordance with international law and the [Draft Articles]" (ILC 2008, art. 3). Given that both international law and the Draft Articles clearly limit state sovereignty, this article seems to uphold and deny sovereignty at the same time. What, if anything, it actually means, is far from clear. Some see in this article a fundamental error in the Draft Articles, namely a focus on sovereignty over and management of the rock in which groundwater is found rather than, as in the UN Convention, on the rights in and management of the groundwater itself (McCaffrey 2009).

The Draft Articles do contain some provisions that recognize the hydro-geologic realities of groundwater. Thus, they require aquifer states to identify recharge and discharge zones of their transboundary aquifers and aquifer systems and to take "special measures" to minimize detrimental impacts of the recharge and discharge processes (ILC 2008, art. 11[1]). The Draft Articles would also extend this duty to non-aquifer states in which a recharge or discharge zone is located (ILC 2008, art. 11[2]). Why non-aquifer states would be willing to do so without sharing the benefits of such actions is nowhere explained. Going beyond the strict needs of working out the relationships of states to internationally shared groundwater, the Draft Articles also would impose an obligation for advanced states to promote capacity building in developing nations (ILC 2008, art. 16). This, like the obligations ostensibly imposed on non-aquifer states regarding recharge and discharge zones, may prove easier to articulate than to enforce.

The Commission recommended in its report that the General Assembly "take note" of the Draft Articles, recommend that States make "appropriate arrangements" for their shared aquifers on the basis of the principles in the Draft Articles, and consider the possibility of convening a negotiating conference to draft a convention on the topic (Yamada 2008). This approach avoids the problem of either suggesting amendments to the UN Convention on the Law of Non-Navigable Uses of International Watercourses or proposing a protocol to that convention. Either approach was troubling given the difficulties the UN Convention has experienced in obtaining the ratifications necessary to enter into force (Salman 2007), although the pace of ratification has picked up recently. On 11 December 2008, the General Assembly adopted a resolution incorporating the suggestions made by the Commission (UNGA 2008). Whether the Draft Articles will ever progress to a treaty, whether as incorporated into a larger treaty on transboundary waters or as a stand-alone instrument, remains entirely uncertain at this point.

Conclusion

The International Law Commission's Draft Articles on the Law of Transboundary Aquifers, having been acknowledged but not approved by the UN General Assembly (UNGA 2008), stand as a significant step in the regularization and codification of the customary international law applicable to internationally shared groundwater. Like so much of the Commission's work, however, it is extremely cautious, effectively neglecting the Commission's charge not only to codify but also to undertake the progressive development of customary international law. States and water users alike will find a useful starting point for working out arrangements for managing at least internationally shared groundwater (or aquifers, if you prefer), but they will quickly discover that satisfactory management will require them to go far beyond the terms of the articles. One of the most obvious points where this will be true is if a recharge zone (or part of a recharge zone) is located outside of an aquifer state (as defined in the articles), satisfactory arrangements will be possible only if the non-aquifer state is accorded a role in managing the aquifer or aquifer system as a whole, including sharing to some extent in the benefits derived from that management, if the non-aquifer state is to be held to the duty allegedly imposed on it by customary international law as set forth in the Draft Articles. In this regard, and perhaps in some other ways, the Berlin Rules on Water Resources provide a more progressive blueprint for the management of internationally shared groundwater.

References

Albiac, J., *et al.*, 2006. The rise and fall of the River Ebro transfer. *Natural Resources Journal*, 46(4), 727–757.

Burchi, S. and Spreij, M., 2003. *Institutions for international freshwater management*. UNESC0 SC-2003/WS/41. Paris: FR.

Burchi, S. and Mechlem, K., eds, 2005. *Groundwater in international law: compilation of treaties and other international instruments*. Study 86. Rome: Food and Agriculture Organization (FAO) Legis.

Burleson, E., 2006. Middle Eastern and North African hydropolitics: from eddies of indecision to emerging international law. *Georgetown International Environmental Law Review*, 18(2), 385–424.

Byers, M., ed., 2000. *The role of law in international politics: essay in international relations and international law*. Oxford: Oxford University Press.

Caponera, D.A. and Alheritiere, D., 1978. Principles of international groundwater law. *Natural Resources Journal*, 18(3), 589–619.

Charturvedi, M.C., 2001. Sustainable development of India's water: some policy issues. *Water Policy*, 3(3), 297–320.

Chavez, O.E., 2000. Mining of internationally shared aquifers: the El Paso-Juarez case. *Natural Resources Journal*, 40(2), 237–251.

Cobbett, P., 1922. *Leading cases on international law*. London: Sweet & Maxwell.

Dellapenna, J.W., 2010. The physical and social bases of quantitative groundwater law. *In*: A. Kelly, ed. *Waters and water rights*. Newark, NJ: LexisNexis: ch. 18.

Eckstein, G.E., 2005. A hydrogeological perspective of the status of ground water resources under the UN watercourse convention. *Columbia Journal of Environmental Law*, 30(2), 525–564.

Eckstein, G.E., 2007. Commentary on the UN International Law Commission's draft articles on the law of transboundary aquifers. *Colorado Journal of International Environmental Law & Policy*, 18(2), 537–610.

Eckstein, G. and Eckstein, Y., 2003. A hydrogeological approach to transboundary ground water resources and international law. *American University International Law Review*, 19(2), 201–258.

EU, 2000. *Water framework directive*, EU Dir. 2000/60/EC, 43 O.J. (L 327). European Union.

Feitelson, E., 2003. When and how would shared aquifers be managed? *Water International*, 28(2), 145–153.

Fuentes, X., 2002. International law-making in the field of sustainable development: the unequal competition between development and the environment. *International Environmental Agreements: Policy, Law & Economics*, 2(1), 109–133.

Gupta, J., 2002. Global sustainable development governance: Institutional challenges from a theoretical perspective. *International Environmental Agreements: Policy, Law & Economics*, 2(2), 361–388.

ILA, 1966. The Helsinki Rules on the uses of the waters of international rivers. *In*: *Report of the Fifty-second Conference* (Helsinki). London: International Law Association, 447–533.

ILA, 1986. The Seoul Rules on international groundwaters. In: *Report of the Sixty-second Conference* (Seoul). London: International Law Association, 251–276.

ILA, 2004. Berlin Rules on Water Resources. *In*: *Report of the Seventy-First Conference* (Berlin). London: International Law Association, 334–421.

ILC, 2006. The Law of Transboundary Aquifers. UN Doc. A/CN.4/L.688 (first reading).

ILC, 2008. The Law of Transboundary Aquifers. UN Doc. A/CN.4/L.724 (2008) (second reading).

Kallis, G. and Butler, D., 2001. The EU water framework directive: measures and implications. *Water Policy*, 3, 125–142.

Kemper, K.E., Mestre, E., and Amore, L., 2003. Management of the Guarani aquifer system: moving towards the future. *Water International*, 28(2), 185–200.

Koh, H.H., 1997. Why do nations obey international law? *Yale Law Journal*, 106(6), 2599–2659.

Kolb, R., 2003. Selected problems in the theory of customary international law. *Netherlands International Law Review*, 50(1), 119–150.

Lepard, B.D., 2010. *Customary international law: a new theory with practical applications*. Cambridge: Cambridge University Press.

McCaffrey, S.C., 2009. The International Law Commission adopts Draft Articles on transboundary aquifers. *American Journal of International Law*, 103(2), 272–293.

Mekong Agreement, 1995. *Agreement on cooperation for the sustainable development of the Mekong River Basin,* (signed 5 April 1995, Cambodia-Laos-Thailand-Vietnam), reprinted in *International Legal Materials*, 34(3), 864–880.

Norman, G. and Trachtman, J.P., 2005. The customary international law game. *American Journal of International Law*, 99(3), 541–580.

Rittberger, V. and Mayer, P., eds, 1993. *Regime theory and international relations*. Oxford: Clarendon Press.

Rubin, A.P., 1977. The international legal effects of unilateral declarations. *American Journal of International Law*, 71(1), 1–30.

Salman, S.M.A., 2007. The United Nations watercourses convention ten years later: why has its entry into force proven difficult? *Water International*, 32(1), 1–15.

UN, 1970. *Progressive Development and Codification of the Rules of International Law Relating to International Watercourses*. [online] G.A. Res. 2669, UN GAOR, 25th Sess., Supp. No. 8, U.N. Doc. A/8028 (1970). Available from: http://daccess-dds-ny.un.org/doc/UNDOC/GEN/NR0/759/52/IMG/NR075952.pdf?OpenElement [Accessed 29 July 2011].

UN, 1997. *Convention on the Law of Non-Navigational Uses of International Watercourses*. [online] G.A. Res. 51/229, UN GAOR, 51st Sess., UN Doc. A/RES/51/229 (1997). Available from: http://internationalwaterlaw.org/documents/intldocs/watercourse_conv.html [Accessed 29 July 2011].

UNGA, 2008. *UN General Assembly Resolution on the Law of Transboundary Aquifers*. A/RES/63/124. Available from: http://internationalwaterlaw.org/documents/intldocs/UNGA_Resolution_on_Law_of_Transboundary_Aquifers.pdf [Accessed 29 July 2011].

Wolfke, K., 1993. *Custom in present international law*. 2nd rev. ed. Dordrecht: M. Nijhoff.

Yamada, C., 2008. *Fifth report on shared natural resources: transboundary aquifers*. UN Doc. A/CN.4/591.

Zeitoun, M., 2007. The conflict vs. cooperation paradox: fighting over or sharing of Palestinian-Israeli groundwater. *Water International*, 32(1), 105–120.

The International Law Commission's flawed Draft Articles on the Law of Transboundary Aquifers: the way forward

Stephen C. McCaffrey

In 2008 the United Nations International Law Commission finalized Draft Articles on the Law of Transboundary Aquifers; they were annexed to a United Nations General Assembly resolution later that year. In Autumn 2011, the General Assembly will decide on the final form of these articles. Unfortunately, the articles overlap with the 1997 United Nations Watercourses Convention and contain an even more serious flaw: they introduce the novel and potentially dangerous concept that a state has sovereignty over the portion of a transboundary aquifer located within its territory. After providing background on the Draft Articles, this paper proposes possible remedies for these problems.

Introduction

In 2008, the United Nations (UN) International Law Commission (ILC) finalized a set of Draft Articles on the Law of Transboundary Aquifers (ILC 2008, McCaffrey 2009, McCaffrey 2010, McCaffrey and Neville 2010). These Draft Articles were annexed to a UN General Assembly (UNGA) resolution (63/124) adopted later that year (UNGA 2008). The General Assembly is to decide in the autumn of 2011 on the final form of the Draft Articles. Unfortunately, the articles overlap with the UN Convention on the Law of the Non-Navigational Uses of International Watercourses (UN 1997). Perhaps more seriously, the Draft Articles introduce the novel and potentially counterproductive concept that a state has sovereignty over the portion of a transboundary aquifer located within its territory. After providing background on the Transboundary Aquifers Draft, this paper will propose a way in which the two problems just identified might be remedied.

The ILC's Transboundary Aquifers Draft

Genesis

The International Law Commission was prompted to take up the study of the Law of Transboundary Aquifers chiefly by the fact that the Commission's earlier Draft Articles on the Law of International Watercourses – which formed the basis of the negotiation of the UN Watercourses Convention 1997 (UN 1997) – did not deal with groundwater that does not interact with surface water. The ILC's articles on watercourses, and the

UN Watercourses Convention, define "watercourse" as: "a system of surface waters and groundwaters constituting by virtue of their physical relationship a unitary whole and normally flowing into a common terminus" (UN Watercourses Convention, Article 2[a].) This definition recognizes that water moves in a "system" consisting of both surface water and groundwater: water may be in a stream now, underground later and emerge still later on the surface again. Most of Earth's fresh water moves in this kind of cycle. It is therefore a mistake to treat surface water and groundwater entirely separately for purposes of legal regulation. This is true whether the issue is one of allocation, pollution, ecosystem protection or management.

However, the ILC realized that its articles on watercourses did not cover groundwater that is not hydrologically connected to surface water. This kind of groundwater is less common; perhaps its best-known form is the "fossil water" found, inter alia, in the Middle East and North Africa. The ILC adopted the term "confined groundwater" to cover such water contained in aquifers that does not relate to surface water. (Whether this is the correct term from a hydrological standpoint – and it may not be – is not material for purposes of the ILC's work, since it is the definition of the term – "groundwater not related to an international watercourse" [ILC 1994] – that controls). When the ILC completed its Draft Articles on International Watercourses in 1994, it adopted a Resolution on Confined Transboundary Groundwater. That resolution states the ILC's "view that the principles contained in its draft articles on the law of the non-navigational uses of international watercourses may be applied to transboundary confined groundwater". But the Commission also recognized "the need for continuing efforts to elaborate rules pertaining to confined transboundary groundwater" (ILC 1994). I have recounted elsewhere the history of how the Commission began the study of transboundary aquifers (McCaffrey 2009), but for present purposes it will suffice to note that it was an outgrowth of the topic, Shared Natural Resources, that the ILC included in its programme of work in 2002. In that year, the Commission also appointed Chusei Yamada as special rapporteur for the topic.

Process

Ambassador Yamada submitted a total of five reports to the ILC on the Law of Transboundary Aquifers, from 2003 to 2008. He was assisted in elaborating the Draft Articles by four working groups established by the Commission. While the special rapporteur chaired the first working group, which is the usual practice, the second through fourth were chaired by Enrique Candioti, a member of the ILC from Argentina. Members of the ILC serve in their individual capacities and not as representatives of their governments. But Argentina has a strong interest in the subject of transboundary groundwater, being one of four states sharing the Guarani Aquifer in central South America. The Guarani is thought to be Earth's largest groundwater resource. An Agreement on the Guarani Aquifer was concluded on 2 August 2010, among Argentina, Brazil, Paraguay and Uruguay (Guarani Agreement 2010). The Agreement provides in Article 2 that the parties exercise sovereign territorial control ("*el dominio territorial soberano*") over their respective portions of the Guarani Aquifer System. This article is comparable to the problematic Article 3 of the ILC's Transboundary Aquifers articles, discussed below. Whether or not Ambassador Candioti was in fact influenced by his country's position vis-à-vis the Guarani, the appearance of possible influence is regrettable because it casts a shadow over the Draft Articles. One would hope that the ILC could avoid such apparent conflicts in the future, as it has generally done with regard to special rapporteurs.

One of the most welcome features of the Commission's work on transboundary aquifers is that the ILC had the benefit of briefings by experts on groundwater and associated geology from several organizations specialized in the field: the United Nations Educational, Scientific and Cultural Organization (UNESCO), the Food and Agriculture Organization of the UN (FAO), the UN Economic Commission for Europe (ECE) and the International Association of Hydrogeologists (IAH). The UNESCO team included a leading expert in the field of international groundwater law, Professor Gabriel Eckstein. Thus the Commission had the benefit of good advice on both the science and the law of transboundary aquifers.

What went wrong?

The overlaps with the UN Watercourses Convention

As indicated above, the Transboundary Aquifers Draft overlaps with the UN Watercourses Convention (McCaffrey 2010). These overlaps are due chiefly to the fact that the physical subject matters of the two instruments are defined in such a way as to cause them to overlap. If the applicable substantive rules of the two instruments were identical, overlapping subject matter coverage would not be a problem. But of course, the substantive rules are not identical. The most dramatic difference is, of course, the "general principle" of sovereignty of aquifer states, which finds no counterpart in the UN Convention. This provision would require a fundamental re-thinking of how basic principles of the Convention – such as equitable and reasonable utilization, prevention of significant harm and prior notification of planned measures – would apply to situations in which both instruments were applicable. Other provisions of the Aquifers Draft also raise questions. For example, the draft, properly, covers recharge and discharge zones, which may not lie in an "aquifer state" but which would generally constitute parts of an "international watercourse" as defined in the UN Convention. Since the Aquifers Draft imposes obligations almost exclusively on aquifer states, those states that are not aquifer states but which contain recharge or discharge zones have only obligations of cooperation (for example, Article 11.2). This seems to be a serious gap in the Aquifers Draft, but it could override the UN Convention if the draft entered into force after the Convention. These are only a few examples of overlaps between the two instruments.

These overlaps between two instruments that originated in the International Law Commission are unfortunate and, to the author's knowledge, unprecedented in the work of the Commission. The ILC ordinarily takes pains to ensure that its work does not overlap with, or otherwise contain inconsistencies with respect to, other work it has done or is currently involved in. Why did it not do so here?

First, the membership of the ILC has changed almost completely since it worked on the international watercourses articles. There may not have been sufficient institutional memory of that work among members to prompt them to avoid overlaps and inconsistencies. While the UN Secretariat is ordinarily the keeper of the organization's institutional memory, including that of bodies such as the ILC, all it can do is call past work to the attention of members of the body in question; it would normally do so, and there is no evidence that it did not do so here.

A second factor may have been a change in the approach of the special rapporteur. At the very outset of the work on shared natural resources, Ambassador Yamada "indicated his intention to deal with confined transboundary groundwaters, oil and gas in the context of the topic and proposed a step-by-step approach beginning with groundwaters" (ILC 2008). But this plan to take up "confined transboundary groundwaters" changed. Ambassador Yamada later stated that: "It would be appropriate to begin with the consideration of groundwaters as the follow-up of the Commission's previous work on the codification of surface waters . . ." (ILC 2006). The reference to "the Commission's previous work on the codification of surface waters" is of course not altogether accurate since as noted above, the ILC's previous work on international watercourses dealt not only with surface water but also with much of the world's groundwater: that which is hydrologically related to surface water. At some point, the special rapporteur seems to have formed the view that his mandate had been broadened, to cover not only "confined transboundary groundwaters", but "groundwaters" in general.

A third factor may have had to do precisely with the excellent scientific advice the special rapporteur received from the organizations indicated above. These experts may have believed that the ILC's work presented an excellent opportunity to codify and progressively develop rules of international law governing all forms of groundwater, not just that which is not covered by the UN Watercourses Convention. They could not be blamed for such a view, since they are not experts in international law and would not have been aware of possible conflicts with the already-existing UN Watercourses Convention. But it is possible that the special rapporteur may himself have been influenced by this kind of thinking. The ILC relies heavily on its special rapporteurs and would generally not question their recommendations – especially if, as suggested above – there was not an adequate institutional memory of the ILC's previous work on international watercourses.

The possibility that the special rapporteur deliberately decided to expand the scope of the transboundary aquifers articles into areas already covered by the UN Watercourses Convention is suggested by a draft article he proposed on the relationship of the aquifers articles to other international agreements:

> *[Article 20] Relation to other conventions and international agreements*
>
> (1) *The present draft articles shall not alter the rights and obligations of the States parties which arise from other conventions and international agreements compatible with the present draft articles and which do not affect the enjoyment by other States parties of their rights or the performance of their obligations under the present draft articles.*
>
> (2) *Notwithstanding the provisions of paragraph 1, when the States parties to the present draft articles are parties also to the Convention on the Law of the Non-navigational Uses of International Watercourses, the provisions of the latter concerning transboundary aquifers or aquifer systems apply only to the extent that they are compatible with those of the present draft articles.* (ILC 2008)

This draft article proved controversial within the Commission and was eventually dropped before the Draft Articles were transmitted to the General Assembly. But it reveals the view of the special rapporteur that the Draft Articles – assuming they were eventually to take the form of a binding treaty, and that the UN Convention had entered into force – should take precedence over the UN Convention to the extent that the two were incompatible. He was thus aware of the possible inconsistencies; rather than trying to eliminate them, the special rapporteur chose to push forward with a new legal regime – a regime that goes well beyond the one originally intended by the Commission in its 1994 Resolution, which would have covered only forms of groundwater not dealt with in the UN Watercourses Convention, and one which in fact overlaps with the Convention.

At a conference on transboundary aquifers in December 2010, sponsored by UNESCO, both Ambassador Yamada and Ambassador Candioti stated that overlaps and inconsistencies as between the UN Watercourses Convention and the ILC's Transboundary Aquifers Articles did not pose a problem because they would be governed by the law of treaties.

Leaving aside the fact that this solution simply invites disputes between states as to, for example, whether there is a conflict, the extent to which the later treaty is compatible with the earlier one (the general rule of treaty law under Article 30[3] of the Vienna Convention on the Law of Treaties is that the later treaty applies to the extent of incompatibility), and so on, it will be of small comfort to water managers, who must have clear direction as to the law to be followed regarding the management of transboundary aquifers and international watercourses. In addition, assuming the aquifers articles eventually become a treaty, it is not clear whether that treaty or the UN Watercourses Convention would enter into force first. If a transboundary-aquifers treaty became effective before the watercourses convention, significant portions of the aquifers treaty could be trumped.

The inclusion of the concept of sovereignty over shared water resources

The first "general principle" contained in the ILC's Transboundary Aquifers Articles is the following:

> *Article 3*
> *Sovereignty of Aquifer States*
> *Each aquifer State has sovereignty over the portion of a transboundary aquifer or aquifer system located within its territory. It shall exercise its sovereignty in accordance with international law and the present articles.*

In this one provision, the ILC has managed to reverse over 100 years of development of international-watercourse law. For it was in 1895 that US Attorney-General Judson Harmon issued his now infamous opinion that has become known as the "Harmon Doctrine" – a doctrine that will be familiar to many readers of this journal (Harmon 1895). In his opinion, Attorney-General Harmon stated in part: "The fundamental principle of international law is the absolute sovereignty of every nation, as against all others, within its own territory . . . The immediate as well as the possible consequences of the right asserted by Mexico [to a portion of the waters of the Rio Grande, which it shares with the United States] show that its recognition is entirely inconsistent with the sovereignty of the United States over its national domain . . ." (Harmon 1895). The United States has never followed the Harmon Doctrine in its relations with co-riparian states and in fact resolved the dispute with Mexico that gave rise to the doctrine in a 1906 treaty entitled Convention Concerning the Equitable Distribution of the Waters of the Rio Grande for Irrigation Purposes.

Despite the fact that it is to be exercised in accordance with international law and the Draft Articles, the draft's conferral of "sovereignty" upon a state as to the portion of a transboundary aquifer located in its territory risks encouraging that state to drill first and ask questions later – or, more likely, to wait to see if its neighbour asks questions later. Article 3 conflates a state's undoubted title (or sovereignty) to the geologic formation within its territory, on the one hand, with the water contained in that formation, on the other. A state may be said to have "sovereignty" over the rock, but not over the water, which is shared with the other state or states overlying the aquifer. A state simply cannot have the exclusive ownership that sovereignty implies in something that is shared with another state. In fact, as discussed elsewhere, the entire concept of "sovereignty" in international relations is highly questionable and tends to be used as a fig leaf to cover up ill-advised, improper or unlawful conduct (McCaffrey 2009, McCaffrey 2010, McCaffrey and Neville 2010, Henkin 1994). In any event, the concept is not a part of customary international watercourse law (International Court of Justice 1997, McCaffrey 2007).

The way forward

The foregoing analysis begs the question whether the ILC's Transboundary Aquifers Draft is salvageable. I believe that it is, but only if: (1) the sovereignty principle is excised; and (2) the draft is given the form of a "practice guide", or the like. I have made related proposals before (McCaffrey 2010).

Excision of the sovereignty principle would be simple: Article 3 could simply be deleted. Alternatively, I have proposed converting it into an article on "Cooperation Between Groundwater States", which would do two things: it would shift the focus to cooperation, where it belongs; and it would replace the legally troublesome concept of "aquifers" with "groundwater" – the latter throughout the Draft Articles (McCaffrey 2010).

Curing the overlaps with the UN Watercourses Convention would be far more complex. It could be done, but would require restricting the scope of the draft *ratione materiae* to groundwater that does not interact with surface water – what the ILC referred to in its 1994 Resolution as "confined" transboundary groundwater. In addition to the complexity of the task, many of the valuable management lessons contained in the Draft Articles would be lost, precisely because their scope would be limited to groundwater that does not receive significant recharge from the surface.

In order to preserve those lessons, the Draft Articles – with a revised or deleted Article 3 – could take the form of a guide to practice. This is the intended form of the ILC's Draft Articles on the highly important and complex subject of Reservations to Treaties (for example, ILC [2010]) and has been so practically since the outset of the ILC's work on the topic over 15 years ago. Similarly, the ILC's 2001 Draft Articles on State Responsibility ("Responsibility of States for Internationally Wrongful Acts"), one of the most fundamental fields of international law, have not been used as the basis of negotiating a convention (ILC 2001). Those draft articles have nevertheless been invaluable to states, scholars and others as a codification of the rules of international law on the subject. In addition to this precedent, there is always the risk that if a treaty is negotiated on the basis of an ILC draft, important aspects of the draft might be lost in the negotiations because they touch on political nerves, might bear upon ongoing disputes, or for other reasons. Thus it is recommended that a working group be formed within the Sixth (Legal) Committee of the General Assembly to make the required adjustments to the ILC's Transboundary Aquifers Draft, converting it into a guide to practice. That guide could then be adopted by the General Assembly as an annex to a resolution.

Conclusion

The International Law Commission's Draft Articles on the Law of Transboundary Aquifers have the potential to make a constructive contribution to the law and practice in the field if the kinds of adjustments recommended in the previous section are made. On the other hand, if the General Assembly decides in the fall of 2011 to convene a conference for the purpose of negotiating a convention on the basis of the ILC's Draft Articles, a valuable opportunity will have been lost. Let us hope that the General Assembly, in its wisdom, will make the right decision.

References

Guarani Agreement, 2010. *Agreement on the Guarani Aquifer between Argentina, Brazil, Paraguay and Uruguay* (2010). Available from: http://www.internationalwaterlaw.org/documents/regionaldocs/Guarani_Aquifer_Agreement-Spanish.pdf [accessed 19 July 2011]. (in Spanish)

Harmon, J., 1895. *Official opinions of the attorneys-general of the United States*, vol. 21, p. 281.
Henkin, L., 1994. The mythology of sovereignty. *In*: R.S.J. Macdonald, ed. *Essays in honour of Wang Tieya*. London: Martinus Nijhoff Publishers.
ILC, 1994. *Resolution on Transboundary Groundwater*. 1994, ILC Report, 2 Y.B. Int'l L. Comm'n, pt. 2, p. 135, UN Doc. A/49/10 (1994).
ILC, 2001. *Report of the International Law Commission on the work of its fifty-third session*, UN GAOR, 56th Sess., Supp. No. 10, UN Doc. A/56/10 (2001) [online]. Available from: http://www.un.org/law/ilc/ [Accessed 29 July 2011].
ILC, 2006. *Report of the International Law Commission on the work of its fifty-eighth session*, UN GAOR, 61st Sess., Supp. No. 10, UN Doc. A/61/10 (2006) [online]. Available from: http://www.un.org/law/ilc/ [Accessed 29 July 2011].
ILC, 2008. Draft Articles on the Law of Transboundary Aquifers, 2008. ILC Report, UN GAOR, 62d Sess., Supp. No. 10, UN Doc. A/63/10 (2008). Available from: http://www.un.org/law/ilc/
ILC, 2010. *Report of the International Law Commission on the work of its sixty-second session*, UN GAOR, 65th Sess., Supp. No. 10, UN Doc. A/65/10 (2010) [online]. Available from: http://www.un.org/law.ilc. [Accessed 29 July 2011].
International Court of Justice, 1997. *Case concerning the Gabčikovo-Nagymaros Project. Judgment of 25 September 1997*. [online]. Available from: http://www.icj-cij.org/docket/files/92/7375.pdf [Accessed 29 July 2011].
McCaffrey, S., 2007. *The law of international watercourses*. Oxford: Oxford University Press.
McCaffrey, S., 2009. The International Law Commission adopts Draft Articles on Transboundary Aquifers. *American Journal of International Law*, 103 (2), 272–293.
McCaffrey, S., 2010. Sovereignty and cooperative management of shared water resources in a time of shrinking availability: the role of international law. Paper presented at UNESCO conference on "Transboundary Aquifers: Challenges and New Directions" (ISARM 2010), 6–8 December, 2010, Paris.
McCaffrey, S. and Neville, K., 2010. The politics of sharing water: international law, sovereignty, and transboundary rivers and aquifers. *In*: K. Wegerich and J. Warner, eds. *The politics of water: a survey*. London and New York: Routledge, 18–44.
UN, 1997. *Convention on the Law of Non-Navigational Uses of International Watercourses*. [online] G.A. Res. 51/229, UN GAOR, 51st Sess., UN Doc. A/RES/51/229 (1997). Available from: http://internationalwaterlaw.org/documents/intldocs/watercourse_conv.html [Accessed 29 July 2011].
UNGA, 2008. UN General Assembly Resolution on the Law of Transboundary Aquifers. [online]. A/RES/63/124. Available from: http://internationalwaterlaw.org/documents/intldocs/UNGA_Resolution_on_Law_of_Transboundary_Aquifers.pdf [Accessed 29 July 2011].

The World Bank policy and practice for projects affecting shared aquifers

Salman M.A. Salman

The policies issued by the World Bank since 1956 for regulating its operations on international waterways dealt only with surface water, and did not make any reference to groundwater. Nevertheless, since 1990, the World Bank addressed, through practice and precedents, projects involving shared aquifers in the same manner it has been dealing with shared surface waters. The article discusses how the Bank reached and applied such an approach, and analyzes its contribution to the evolution and progressive development of the law involving transboundary groundwater.

Introduction

The World Bank (hereinafter the Bank) faced the challenges emanating from its funding of projects on international watercourses a few years after its establishment in 1946. As an international financial cooperative institution, the Bank realized from the start the delicate and complex nature of such projects. Indeed, its Articles of Agreement of themselves require the Bank to act prudently in the interests both of the particular member that is benefiting from the project, as well as the other members of the institution as a whole (World Bank 1946). Those challenges were compounded by the paucity of principles governing international watercourses at that time, the large number of shared watercourses and the existence of disputes on some of them. Nonetheless, the Bank was able to develop some approaches for dealing with projects on international watercourses. Such approaches included not financing projects that could cause appreciable harm to any of the other riparians, or where a dispute existed between the beneficiary state and any of the riparians, unless an agreement was reached thereon. Those general principles gradually evolved, and guided the operations of the Bank on international watercourses during the first decade after it started operational work.

However, in 1956 the Bank was faced, for the first time, with an objection for a project it planned to finance on an international river. Turkey objected to the Ghab irrigation project in Syria, on the Orontes River that the two countries and Lebanon share, because of its concerns that the proposed project would decrease the flow of the Orontes River and cause appreciable harm to its interests. The objection prompted a thorough discussion that resulted in the issuance in 1956 of the first Bank policy for "Projects on International Inland Waterways". The policy established an early-warning system by instructing staff that the Bank's management should be informed promptly of any project the carrying out of which would involve the use of an international inland waterway. The policy went on to state that no steps were to be taken to investigate the merits of the project or to process it without prior approval by management of a procedure for dealing with the international aspects of the project. International inland waterways were defined by the policy to include any river, canal, lake or other inland waterway that forms the boundary between, or flows through two or more states, whether members of the Bank or not. Thus, only surface water was specified in the policy, and the application of the policy was extended to all the riparians of the waterway in question. The latter is a basic requirement of international law that the Bank could not ignore.

The policy was revised and reissued in 1965 to include semi-enclosed coastal waters, namely bays, gulfs, straits or channels bounded by two or more states, or if within one state, recognized as necessary channels of communication between the open sea and other states. The concern was about the likelihood of pollution of such coastal waters as a result of a Bank-financed project. It is worth adding in this connection that the Bank had no policy on the environment in 1965. Thus, any likely negative effects of Bank-financed projects on semi-enclosed waters were addressed through the 1965 policy on international waterways. This situation continues despite the many detailed policies and guidelines on the environment that the Bank has issued since 1982. It should be added that although neither the 1956 nor the 1965 versions of the policy mentioned the requirement of notification of the other riparians of the proposed projects, the Bank followed this requirement, by and large.

In 1985 an elaborate policy for projects on international waterways was issued by the Bank. The policy was issued, again, following an objection, this time by Iran, to the Igdir-Aksu Irrigation Project in Turkey. The Project abstracted water from the Aras River which is shared by the two countries, as well as the Soviet Union at that time. The objection by Iran held back the processing of the project for more than a year, and prompted the Executive Directors of the Bank to instruct Bank management to prepare a more detailed policy dealing with such projects. This was accomplished a year later in April 1985. The 1985 version of the policy was reissued in 1994, and again in 2001, as Operational Policy (OP), and Bank Procedures (BP) 7.50, "Projects on International Waterways". The OP and the BP reiterated, by and large, the same parameters of the 1985 policy (Salman 2009). The 2001 version (the Policy) continues to govern the operations of the Bank on international waterways.

Parameters of the current policy

The Policy maintained the definition of the term "international waterway" which includes shared surface waters as well as semi-enclosed coastal waters. Such waterways include: (1) any river, canal, lake or similar body of water that forms a boundary between, or any river or body of surface water that flows through, two or more states, whether Bank members or not; (2) any tributary or other body of surface water that is a component of any waterway described in (1) above; and (3) any bay, gulf, strait or channel bounded by two or more states or, if within one state, recognized as a necessary channel of communication between the open sea and other states – and any river flowing into such waters.

The Policy does not make a reference to groundwater. The committee that prepared the policy in 1985 considered the issue of transboundary groundwater, but concluded that insufficient technical knowledge and information was available to assist with the inclusion of shared groundwater (Krishna and Salman 1999, p. 193).

The Policy states that cooperation and goodwill of all the riparians is essential for the efficient utilization and protection of the waterway. It goes on to indicate that the Bank attaches great importance to the riparians making appropriate agreements or arrangements for this purpose, and declares the readiness of the Bank to assist the riparians achieve this end. The projects covered and enumerated in the Policy include hydroelectric, irrigation, flood control, navigation, drainage, water and sewerage, industrial and similar projects that involve the use or potential pollution of international waterways. They also include operations for financing detailed design and engineering studies for such projects.

With regard to such operations, the Policy stipulates that the international aspects of a project on an international waterway are dealt with at the earliest possible opportunity in the project cycle. The prospective borrower should notify the other riparians of the proposed project and its details. The notification contains, to the extent available, sufficient technical specifications, information and other data to enable the other riparians to determine as accurately as possible whether the proposed project has the potential for causing appreciable harm through water deprivation or through pollution. The notified riparian states are given a reasonable period, which could range from two to six months (depending on the nature and complexity of the project), to reply to the notification.

If the prospective borrower indicates to the Bank that it does not wish to give notification, normally the Bank itself does so on behalf of the borrower. If the prospective borrower does not wish to give notification, and objects to the Bank's doing so, the Bank discontinues processing of the project. Responses to the notification have varied from giving no objection to the proposed project, to asking for more information and time, to simply not responding to the notification, or objecting to the project. Objections to Bank-financed projects have been rare, and were received in few projects because the Bank would usually ensure that the projects it intends to finance would not cause appreciable harm to the interests of other riparians. This is usually clarified and confirmed in the notification letters.

In case of an objection to the project by one or more of the riparian states, Bank staff assess the objection raised and ascertain whether the proposed project would cause appreciable harm to the other riparians. The staff also makes a recommendation to Bank management on how to proceed. The Bank may, in appropriate cases, appoint one or more independent experts to examine the objection. Such experts have no decision-making role in project processing, and their opinion is submitted for Bank purposes only.[1] Should the Bank decide to proceed with the project despite the objection, the Bank informs the objecting riparian (or riparians) of its decision. The Policy includes three exceptions to the notification requirement, namely: (1) rehabilitation of existing schemes which would not result in appreciable harm to any of the riparians; (2) water-resources surveys and feasibility studies; and (3) projects in a tributary in the lowest downstream riparian, where the tributary runs exclusively within that state, unless there is concern that the project could cause appreciable harm to other states.

Thus, the Bank was able from its very early years to develop a policy for dealing with projects on international waterways. The policy evolved gradually, and the large number of projects that the Bank financed on international waterways provided the opportunity for interpreting many of the provisions, and for filling the gaps therein. This is particularly true about transboundary groundwater, as will be discussed in the next section of this article.

The Bank and projects affecting transboundary groundwater

The Bank faced the issue and the challenges of transboundary groundwater for the first time in 1990, five years after it adopted the 1985 policy. The project that raised the issue of transboundary groundwater was the Sahara Regional Development Project in Algeria. Prior to 1990, the Bank had financed a number of groundwater projects in some countries including Yemen, Tunisia, Jordan, Mexico, Nepal and Bangladesh, but "the complexity of the issues of shared groundwater resources did not arise presumably because either the possibility of whether the concerned aquifers might be shared with neighboring states was not entertained, or simply because it was assumed the aquifers concerned were purely national" (Krishna and Salman 1999, p. 180).

The Sahara Regional Development Project involved the utilization of waters by Algeria from two aquifers (known generally at that time as the Sahara Aquifers) through the drilling of wells for irrigation and water-supply purposes, and provision of related equipment. One of the aquifers, known as the Continental Intercalaire (CI) holds fossil, non-renewable waters. The other aquifer which was known as the Complexe Terminal (CT) is a rechargeable aquifer. While it has been established by geological studies that these aquifers extend from Algeria to Tunisia, the possibility of their extending further east to Libya, and west to Morocco, was indicated by diagrams in some reports. An area of about 800,000 cubic kilometres of these aquifers was studied by the United Nations Educational, Scientific and Cultural Organization (UNESCO) during the period 1968–71, with funding from the United Nations Development Programme (UNDP), as part of a strategy to explore water resources in the desert areas of the world. The study recommended, among other things, that the pumping of the CT should not exceed 60 metres until 2010, and that extraction by each country in either aquifer should have the same impact on the other. Both the Algerian and Tunisian sides endorsed the study and recommendations in 1989.

The Project documents issued in May 1990 mentioned, for the first time, the possibility of "riparian issues" in the proposal that the Bank should seek commitment from the government of Algeria to adhere to the recommendations of the UNDP study for the use of the common groundwater resources. The deliberations within the Bank raised the facts that: (1) the Bank Policy does not expressly cover groundwater; (2) the UNDP study would ensure the best use of the common resources and protection of the common interests of both countries; and (3) the Commission Technique Tuniso/Algerienne pour l'Hydraulique et l'Environnement (CTHE) that the two countries set up should serve as the appropriate forum for the discussion of the issues raised by the proposed project.

However, at that point the issue of applying the Bank Policy for Projects on International Waterways to the Project started to come under serious consideration, based on the fact that the water in question was indeed shared by two states. After more debate, it was decided that Algeria should notify Tunisia of the Project, and that such notification should: (1) indicate the Algerian government's endorsement of the UNDP study, and understanding of a similar endorsement on the part of the Tunisian government; (2) emphasize the measures that the Algerian government plans to take to prevent pollution of the aquifers, and its expectation for similar measures on the part of the Tunisian government; (3) request comments on the Project from the Tunisian side; and (4) give Tunisia a period of not less than one month to respond.

Since the extension of the aquifers to Libya and Morocco was not supported by actual geological studies at that time, and since ascertaining such an extension would have been very difficult and costly, notifying these two countries of the proposed project was considered unwarranted. The possibility of the World Bank notifying Tunisia on behalf of Algeria was also considered in case Algeria asked the Bank to do so on its behalf. After corresponding with Algeria, the Bank, on 24 October 1990, notified the Tunisian authorities of the Project. The notification included a summary of the Project and its impact on the shared aquifer, as well as the Project Evaluation Report. The notification emphasized the Bank's understanding of the existence of cooperative arrangements between the two countries regarding the aquifers, and their intention to adhere to the UNDP study recommendations. Tunisia was given one month to respond.

No response was received from Tunisia. The Bank proceeded with the processing of the Project, and the Project was presented to the Executive Directors of the Bank on 24 September 1991. It is worth adding that the legal documents included a covenant stating that: "The Borrower shall take all measures necessary to ensure that the use of the groundwater resources in [the Northern Sahara] Region shall be planned in accordance with the recommendation of the Study of Water Resources in the Sahara."[2]

Thus, the Bank policy for projects on international waterways which refers only to surface water was extended for the first time to transboundary groundwater. Accordingly, a precedent was established in 1990. It did not differentiate between fossil and replenishable aquifers, or between aquifers connected to surface waters and others. The precedent, no doubt, rests on solid legal basis. The rationale for notification for surface water applies equally to groundwater. Indeed it may even be argued that the rationale for notification applies more strongly in the case of groundwater. This is because of its invisibility and susceptibility to externalities. Once an aquifer is polluted, unlike surface water, it is extremely difficult to clean it up. Once a fossil aquifer is depleted, it is difficult to replenish it to its natural level. The Bank has been acting in the best interests of all riparians in case of surface water, and needs to act the same way in the case of groundwater.

Moreover, the Bank took note of the fact that the International Law Association (ILA) 1966 Helsinki Rules (ILA 1966), as well as its 1986 Seoul Rules (ILA 1968), both addressed transboundary groundwater (Salman 2007). The Helsinki Rules dealt with aquifers connected to surface waters only,[3] and addressed a wide range of issues including procedures for prevention and settlement of disputes, of which notification is a major part (Bogdanovic 2001). The limitation of the Helsinki Rules to those types of aquifers was addressed by the Seoul Rules that extended application of the Helsinki Rules to all types of aquifers, whether connected to surface waters or not.[4] Although the work of the ILA is not legally binding per se, the Helsinki Rules were considered as reflecting customary international law (Bourne 1997), and have thus strengthened the basis for the Bank's approach for transboundary groundwater in the Algeria Project.

In 1993, a year after the Project was approved, the Bank acknowledged more clearly and widely the international character and complexities of shared aquifers. The World Bank *Policy paper on water resources* stated that: "International water issues also involve important groundwater resources. In a number of cases, aquifers cross international boundaries; thus pumping by one country interferes with another country's pumping or stream flows" (World Bank 1993, p. 39).

The Bank faced the issue of transboundary groundwater again in 1997, a few years after the Sahara Regional Development Project. The issue was faced again in the Middle East and North Africa region; this time it was the Disi-Amman Water Conveyor Project in Jordan. Under the proposed Project, Jordan would convey water from the Disi Aquifer to Amman through a 350-kilometre conveyor. A build, operate and transfer (BOT) contract was proposed for which the Bank was expected to offer to the bidders a partial-risk guarantee that would protect commercial lenders against government default (Macoun and El-Naser 1999). The Disi Aquifer is a fossil, non-renewable aquifer which underlies both northern Saudi Arabia and southern Jordan. However, no agreement or joint-management mechanism for the aquifer existed between the two countries at that time. As Greg Shapland observed: "There is a perceptible flow from the Saudi side into Jordan, with water from the aquifer eventually emerging at the surface near the Dead Sea. The reserves are large, but, because much of it is more than 250 meters below ground (the conventional economic pumping limit), it may not be cost effective to abstract more than a small proportion" (Shapland 1997, p. 144). The proposed Project was designed to abstract up to 100 million cubic metres a year (MCM/yr), with the likelihood that the figure may climb to 150 MCM/yr in future to meet the growing demand for water in Amman. The well-fields to be drilled would be about 35 kilometres from the borders with Saudi Arabia.

Although the Bank took note of the fact that OP/BP 7.50 refers only to surface water, it decided to follow the precedent of the Sahara Regional Development Project in Algeria. Moreover, the Bank took note of the adoption by the United Nations General Assembly that same year (1997) by a large majority of the United Nations Convention on the Law of the Non-Navigational Uses of International Watercourses (Watercourses Convention, or the Convention). Although the Convention addressed shared aquifers, application of the Convention to such shared aquifers, similar to the Helsinki Rules, has its limitations, and applies only to groundwater connected to surface water.[5] The Bank also took note of the fact that a number of regional treaties such as the 1992 United Nations Economic Commission for Europe (UN/ECE) Helsinki Convention[6] and the 1995 Southern African Development Community Protocol on Shared Watercourses[7] address transboundary groundwater. Based on the Bank's own precedent and those developments on international water law, the Bank informed Jordan of the need to notify Saudi Arabia of the Project. The Risk Assessment indicated that expected abstractions in Saudi Arabia would not significantly affect water levels in the Jordanian well-fields, but that any new and unexpected developments close to the border represented a potential project risk. Jordan agreed and notified Saudi Arabia of the Project in February 1998. Saudi Arabia requested more information, and later gave a qualified no objection, asking that no wells be drilled within 15 kilometres of the borders of each country. However, processing of the Project was discontinued by the Bank because of issues concerning the partial-risk guarantee.

The Project followed and strengthened the precedent set by the earlier project in Algeria of inclusion of transboundary groundwater in the scope of international waterways covered under the Policy (notwithstanding the absence of explicit reference to such groundwater), and of requiring notification of other riparians of projects that may affect the shared aquifer. It is worth adding that those two projects were in the Middle East and North Africa; a region characterized by water scarcity and disputes.

Those two projects generated wide and extensive discussion within the Bank as to whether the Bank should be financing projects that affect the waters of fossil non-renewable aquifers because of the issue of sustainability of such aquifers. The World Bank Operational Policy on Water Resources Management (OP 4.07) states in Paragraph 2(d) that the Bank assists borrowers in "restoring and preserving aquatic systems and guarding against over-exploitation of groundwater resources, giving priority to the provision of adequate water and sanitation services for the poor". One important factor to be considered by the Bank regarding the question of whether to finance projects that affect fossil aquifers is the project alternatives. Paragraph 2 of OP 4.01, Environmental Assessment (EA) states that: "EA is a process whose breadth, depth, and type of analysis depend on the nature, scale, and potential environmental impact of the proposed project. EA evaluates a project's potential environmental risks and impacts in its area of influence; examines project alternatives . . . " The analysis of alternatives under this OP compares feasible alternatives to the proposed project site, technology, design, and operation – including the "without project" situation. Thus, the issue of project alternatives plays a major role in the decision.

Subsequent to those two projects, the Bank financed a number of projects involving the use of shared aquifers, and notification of other riparians was undertaken (Salman 2009). One of those projects was the 2003 Second National Fadama Development Project in Nigeria. The project consisted of financing demand-driven micro-projects which would include small-scale irrigation and water-supply projects, using both surface and groundwater in the Niger and Lake Chad basins. Nigeria notified all the riparian states of the two basins, as well as the Lake Chad Basin Commission (LCBC)[8] and the Niger Basin Authority (NBA).[9] No concerns were expressed by any of the riparians, the LCBC or the NBA, and the Bank proceeded to finance the Project. Nigeria followed a similar approach for the Fadama III Project in 2007. Unlike the previous project, Fadama III involved only the abstraction of groundwater in the Niger Basin catchment area, and no surface water was involved. After notification by Nigeria, the NBA gave a positive response.

Similarly, three projects in Ethiopia in 2004 involved the use of both surface and ground waters shared with other countries. Those projects are: (1) Water Supply and Sanitation Project; (2) Decentralized Infrastructure Project; and (3) Productive Safety Net Program. The surface waters under the projects related to the Nile and the Wabe Shabelle and Ogaden basins, while the groundwater related to the Rift Valley lakes. All the Nile riparian countries (of both the Blue Nile and the White Nile), as well as Somalia[10] and Djibouti, were notified by the Bank of the project, and no unfavourable responses were received from any of them.

This trend of dealing with shared surface and ground water in the same project continued, and the Inland Waters Project in Croatia in 2007 followed the same approach. Under the project, new well-fields would be developed in the Danube River catchment area within Croatia. Although groundwater extraction under the project would be well below the threshold of ten million cubic metres annually specified in the Espoo Convention as the basis for notification,[11] the Danube River Commission was notified by the Bank, on behalf of Croatia, of the project, and the Commission provided a positive response. It is worth noting that in this Project the Bank went beyond the requirements of the Espoo Convention and applied its own Policy which is more strict than the Convention. Conversely, the Bank would have applied the requirements of a treaty or a convention if such requirements are more strict than those of the Bank Policy.[12] This approach is intended as a confirmation that the project would not cause appreciable harm. The projects discussed above confirmed and elaborated the application of the Bank Policy to all types of aquifers.

The Sahara Aquifers were revisited again, this time in a project in Tunisia in 2009, almost 20 years after the first project. Major developments have taken place during those years on those aquifers. It was determined in the mid-1990s that the two aquifers (CI and CT) are indeed one aquifer, and that Libya also shares this aquifer with Algeria and Tunisia. The aquifer is now called the North Western Sahara Aquifer System (NWSAS), although it is commonly referred to as SASS, the French acronym for Système Aquifère du Sahara Septentrional. The aquifer covers an area of over one million square kilometres; 700,000 in Algeria, 250,000 in Libya and 80,000 in Tunisia. The three parties met in Rome on 19 and 20 December 2002, in the context of a project sponsored by the Food and Agriculture Organization (FAO) for this aquifer, and recorded their agreement for the establishment of a Consultation Mechanism for the Aquifer. The Consultation Mechanism, intended to coordinate, promote and facilitate the rational management of the aquifer, was inaugurated in June 2008 (FAO 2005), under the auspices of Observatoire du Sahara et Sahel (OSS)[13], following the signature of the Declaration by the Ministers of Water Resources of Algeria, Libya and Tunisia for the North-Western Sahara Aquifer System. The declaration stated that the main objective of the Mechanism is to serve as a framework for cooperation and exchange of information among the three riparian states. Since its establishment, the Mechanism has held a series of workshops and meetings in the three countries sharing the aquifer.

The Tunisia Second Water Sector Investment Project included abstraction of water for irrigation in oases covering about 300 hectares from the NWSAS. It also included three deep boreholes to improve knowledge of the aquifer, based on hydrogeological and geophysical studies. Although those wells would fall under the exception to the notification requirement regarding "water resources surveys and feasibility studies" discussed earlier, it was decided that transparency and full disclosure require mentioning them in the notification letter, together with the other components of the project.

With the Consultation Mechanism fully established within OSS, it was agreed that Tunisia could send the notification letter to the Mechanism for forwarding to both Algeria and Libya.[14] No unfavourable response was received from either country, and the Bank proceeded with the processing of the project.

Transboundary groundwater was also considered in connection with the exceptions to the notification requirement under the Bank Policy. One water supply and sanitation project in Turkey (Cesme-Alacati Water Supply and Sewerage Project) included a component dealing with the rehabilitation of the existing water distribution networks in a number of towns. The water supply source under the Project came from groundwater wells and springs. The issues that arose were: (1) whether the wells and springs are connected to any international watercourses; and (2) if so, should those groundwater wells and springs be considered and dealt with as international waterways? In this connection, the Bank was clearly debating whether a domestic aquifer that receives water from an international river should be treated as an international aquifer.[15] Based on the technical information provided, it was concluded that there was no connection between the groundwater wells and springs, and any international waterways. In addition to this finding, it was noted that even if shared surface water was involved, the proposed improvements would be exempt from the notification requirement under the Policy because the project was only financing the rehabilitation of existing schemes. The Project was finally processed under this exception.

Similarly the issue of notification was considered in connection with the Environmental Protection and Sustainable Development of the Guarani Aquifer System Project. The Project involved the four riparian states of the aquifer – Argentina, Brazil, Paraguay and Uruguay – who were all beneficiaries of the project and were participating in its implementation. Accordingly, it was concluded that, because of this reason, the issue of notification was not relevant in the context of this project.

In addition to those projects, the Bank undertook studies in all other projects involving groundwater to ascertain whether the aquifer in question in each project is shared with other states, or is wholly a domestic or national aquifer. The purpose of this exercise of due diligence is simply to ensure that the Bank is indeed acting prudently, in the interests of all the members of the institution, and to determine whether the project is governed by the practice of the Bank that has been established through the precedents discussed above.

This wide application of the Bank practice of including transboundary groundwater as part of the definition of the term "international waterways" has been accepted by the states receiving Bank financing for such projects. Notification to the other states sharing the aquifer in question has been routinely undertaken since 1990, and no challenge to this Bank requirement was raised by any riparian. Through such state practice, the Bank Policy has contributed significantly to the evolution and progressive development of the law of transboundary aquifers.

Conclusion

The discussion in the last section of this article shows that the Bank has recognized transboundary groundwater as a shared resource since 1990, and has dealt with it in the same manner as shared surface water under the Bank Policy. This is notwithstanding the fact that the definition of "international waterways" under the Bank Policy does not make an explicit reference to transboundary groundwater. Indeed, the initial argument made in 1985 about lack of adequate technical knowledge and information as a reason for not including transboundary groundwater in the 1985 Policy did not last long.

The overview of the Bank-financed projects in this area since 1990 indicates that the practice of notifying other riparians of projects that may affect transboundary groundwater is now clearly established. The projects where the Policy was applied ranged from ones that depended exclusively on fossil aquifers, to others where the aquifers are connected to surface waters and thus are rechargeable. The projects spread through most of the regions of the globe, although those affecting shared aquifers in the Middle East and North Africa region have contributed most significantly to the process. Notification under the projects reviewed above has been undertaken in some cases by the Bank at the request of the borrowers, and in others by the borrowers themselves. The notified entities ranged from government ministries to river-basin commissions or aquifer joint-management organizations. The Bank applied its own Policy over requirements of regional treaties when such requirements are more strict than those of the regional treaties.

This process is similar in all aspects to the notification process for surface water under the Bank Policy and procedures. Yet it is based on practice and precedents, rather than an amendment to the policy, to expand the definition of the term "international waterways" to include an explicit mention of transboundary groundwater. Such an amendment, which is currently under consideration within the Bank, would simply codify the existing practice of treating transboundary groundwater in the same manner as shared surface water for all the purposes specified under the Bank Policy for projects on international waterways, particularly with regard to notification. This practice has no doubt widened and strengthened state practice with regard to shared aquifers.

Although it is based on the existing principles of customary international water law, the Bank Policy and practice have made major contributions to the evolution and progressive development of such principles, including those related to shared aquifers. This contribution is evident in the extensive and consistent state practice in the wide range of projects on international waterways, whether affecting surface or ground waters, that the Bank has financed, and where the Bank Policy and practice were followed.

Notes

1. The Bank has so far appointed such experts in only one project in 1989: the Baardhere Dam and Water Infrastructure Project on the Juba River in Somalia. Ethiopia (which shares the river with Somalia) responded to the notification with a detailed and substantiated objection to the Project, claiming appreciable harm to its interests. That objection prompted the Bank to appoint three independent experts to examine the objection. The experts confirmed the Bank's determination that the Project would not cause appreciable harm to Ethiopia's interests. For more details on the project and the objection see Salman (2011).
2. Section 4.02 of the Loan Agreement between Democratic and Popular Republic of Algeria and International Bank for Reconstruction and Development, dated 6 November 1992 (Loan No. 3405-AL).
3. Article II of the Helsinki Rules defined international drainage basin as "a geographical area extending over two or more States determined by the watershed limits of the system of waters, including surface and underground waters, flowing into a common terminus".

4. Article 2(2) of the Seoul Rules stated that "an aquifer intersected by the boundary between two or more States that does not contribute water to, or receive water from, surface waters of international drainage basin constitutes an international drainage basin for the purpose of the Helsinki Rules".
5. Similar to the Helsinki Rules, the UN Watercourses Convention defines the term "watercourse" as "a system of surface waters and groundwaters constituting by virtue of their physical relationship a unitary whole and normally flowing into a common terminus". Thus, the Convention applies only to groundwater that is connected to surface water. To remedy this situation the International Law Commission (ILC), after completing its work on the draft Convention, issued "ILC Resolution on Confined Transboundary Groundwater". The Resolution, like the Seoul Rules, commended the states to be guided by the principles contained in the Draft Convention in regulating transboundary groundwater. For an analysis of how the UN Watercourses Convention addressed groundwater see Eckstein (2005) and McCaffrey (2007). Both authors, as well as other experts, have pointed at the problem of the use of the term "confined aquifers".
6. See Convention on the Protection and Use of Transboundary Watercourses and International Lakes, 1992 . The Convention defines transboundary water to include both surface and groundwaters that mark, cross or are located on boundaries between two or more states.
7. See the Protocol on Shared Watercourse Systems in the Southern African Development Community (SADC) Region, 1995 . The Protocol defined "watercourse system" to mean the inter-related hydrologic components of a drainage basin such as streams, rivers, lakes, canals and underground water which constitute a unitary whole by virtue of their physical relationship." This Protocol was replaced in 2000 by the Revised Protocol on Shared Watercourses in the Southern African Development Community (SADC). The Revised Protocol adopted the definition of the term "watercourses" of the UN Watercourses Convention (see n. 5).
8. It should be noted that the 1964 Convention and Statutes of Lake Chad Basin (Cameroon, Chad, Niger and Nigeria, joined in 1994 by the Central African Republic) made an explicit reference to groundwater when they stated that the exploitation of the Basin " . . . and especially the utilization of the surface and under groundwaters, has the widest meaning and refers in particular to the needs of domestic, industrial and agricultural development". This reference to groundwater, made in 1964, is one of the earliest such references to groundwater in a regional treaty.
9. It is worth noting in this connection that the 1980 Convention Creating the Niger Basin Authority (NBA) (Benin, Cameroon, Chad, Côte d'Ivoire, Guinea, Mali, Niger, Nigeria and Upper Volta) vested the NBA with extensive authority over both surface and groundwater, including "the initiating and monitoring of an orderly and rational regional policy for the utilization of the surface and underground waters in the Basin".
10. Because of the absence of a functional government in Somalia, the notification was sent to the Executive Director representing Somalia in the Board of Directors of the World Bank. It was reconfirmed to him that the Project would not cause any appreciable harm to Somalia's interests.
11. The Espoo Convention (Convention on Environmental Impact Assessment in a Transboundary Context 1991) lists in Appendix 1 a number of activities that require notification of the Secretariat of the Convention by the party planning any such activity. The list includes "groundwater abstraction activities in cases where the annual volume of water to be abstracted amounts to 10 million cubic metres or more".
12. For example, the Bank consistently applies the provisions of treaties and conventions when they require notification or exchange of data and information even if the project in question would fall under one of the three exceptions to the notification requirement under the Bank Policy discussed earlier.
13. The OSS is an independent international organization based in Tunis, Tunisia. It was founded in 1992 to improve early-warning and monitoring systems for agriculture, food security and drought in Africa. Membership of OSS consists of 22 African countries, five countries in Europe and North America (Germany, Canada, France, Italy and Switzerland), and four sub-regional African organizations. For more information on OSS, see: http://www.oss-online.org/index.php?option=com_content&task=view&id=433&Itemid=564&lang=en.
14. By the time of notification of the Mechanism of the Project in early 2009, the United Nations General Assembly had already passed a resolution in December 2008 taking note of the

Articles on the Law of Transboundary Aquifers that were adopted by the ILC in August 2008. For the history and an analysis of the Articles see Eckstein (2007).

15. Julio Barberis addressed this matter in 1986 and concluded that groundwater may become of international relevance in the following situations: (1) where a confined aquifer is intersected by an international boundary; (2) where an aquifer lies entirely within the territory of one state but has interconnections and interdependence with an international watercourse; (3) where the aquifer is entirely situated within the territory of one state but has interconnections and interdependencies with another aquifer in another state; and (4) where an aquifer is entirely situated within the territory of one state but is getting recharged in another state (Barberis 1986). This classification raises the question of whether a river that is situated in one country but has connections to an international aquifer can still be considered a national river. In the author's view, such a river is indeed an international river because of such connections.

References

Barberis, J., 1986. *International groundwater resources law*. Legislative Study 40. Rome: FAO.

Bogdanovic, S., 2001. *International law of water resources. Contribution of the international law association* (1954–2000). The Hague: Kluwer Law International.

Bourne, C., 1997. *International water law. Selected writings of Professor Charles Bourne*. The Hague: Kluwer Law International.

Eckstein, G., 2005. A hydrological perspective of the status of ground water resources under the UN Watercourse Convention. *Columbia Journal of Environmental Law*, 30(4), 525–540.

Eckstein, G., 2007. Commentary on the U.N. International Law Commission's Draft Articles on the Law of Transboundary Aquifers. *Columbia Journal of International Environmental Law and Policy*, 18(4), 537–552.

FAO, 2005. *Groundwater in international law. Compilation of treaties and other international instruments*. Legislative Study 86. Rome: FAO.

ILA, 1966. *International Law Association. Report of the Fifty-Second Conference*, Helsinki. London: ILA.

ILA, 1968. *International Law Association. Report of the Fifty-Fourth Conference,* Seoul. London: ILA.

Krishna, R. and Salman, S., 1999. International groundwater law and the World Bank Policy for Projects on Transboundary Groundwater. *In*: Salman M.A. Salman, ed. *Groundwater: legal and policy perspectives*. World Bank technical paper 456. Washington, DC: The World Bank.

McCaffrey, S., 2007. *The law of international watercourses*. 2nd ed. New York: Oxford University Press.

Macoun, A. and El Naser, H., 1999. *Groundwater resources management in Jordan: policy and regulatory issues*. *In*: Salman M. A. Salman, ed. *Groundwater, legal and policy perspectives*. World Bank Technical Paper 456. Washington, DC: World Bank.

Salman, S., 2007. The Helsinki Rules, the United Nations Watercourses Convention and the Berlin Rules. Perspectives on international water law. *International Journal of Water Resources Development*, 23 (4), 625–640.

Salman, S., 2009. *The World Bank Policy for Projects on International Waterways. An historical and legal analysis*. Washington, DC: World Bank and Leiden: Martinus Nijhoff.

Salman, S., 2011. The Baardhere Dam and Water Infrastructure Project in Somalia. Ethiopia's objection and World Bank response. *Hydrological Sciences Journal*, 56(4), 630–640.

Shapland, G., 1997. *Rivers of discord. International water disputes in the Middle East*. London: Hurst & Company.

World Bank, 1946. *International Bank for Reconstruction and Development. Articles of agreement* [online]. Article III, Section 4(v), (Washington DC). Available from: http://web.worldbank.org/WBSITE/EXTERNAL/EXTABOUTUS/0,,contentMDK:20049557~menuPK:63000601~pagePK:34542~piPK:36600~theSitePK:29708,00.html[Accessed 15 May 2011].

World Bank, 1993. *The World Bank water resources management policy paper*. Washington, DC: World Bank.

Governance of transboundary aquifers: new challenges and new opportunities

Jamie Linton and David B. Brooks

This paper discusses the emerging role of water governance as applied to transboundary aquifers. To allow for important differences in governance principles from transboundary surface water, the authors suggest that greater attention be paid to the scale of the aquifer when developing interest-articulation and decision-making processes. The authors further introduce five objectives that, if met, would help realize equitable and reasonable use of these aquifers. Good governance can best be met through appropriate scales of interest articulation and decision making, and involvement of a broad range of non-state actors as well as formal agencies of the state.

Introduction

It is time for the global water community to insist upon improved governance for transboundary aquifers. The problems resulting from excessive and competitive pumping of aquifers and of uncontrolled land use above aquifers are becoming so evident that new modes of cooperation and shared decision making are imperative. Puri and Aureli (2005, p. 661) assert:

> Transboundary aquifers are as important a component of global water resource systems as are transboundary rivers; yet their recognition in international water policy and legislation is very limited.

The governance of transboundary aquifers implies new challenges and new opportunities, and the main purpose of this paper is to explore some of those challenges and opportunities. Issues arising from over-consumption and contamination, which occur with many aquifers, are "exacerbated [with transboundary aquifers] by the difficulty of ensuring the ability of all stakeholders to articulate their interests, and the difficulty of effectively coordinating management", which are among the key goals of good governance (Norman and Bakker 2004, p. 3). Specifically, we consider how international aquifers constitute a scale at which decision-making processes take place (or might take place) as well as how participatory decision making can articulate with the broad objectives of sustainable management of international groundwater resources.

Our argument may be considered an elaboration of the position of Burke *et al.* (1999) and Burke and Moench (2000) that groundwater management has for too long been seen as a technical issue when the main constraints are social, institutional and political. It also fits comfortably with the thesis of Linton (2010) that water has to be seen as a social fact as well as a physical substance, and with that of Espelend (1998) that the political dimensions of water need to be considered as well as the scientific and economic dimensions. Though Burke and Moench do not apply their arguments specifically to transboundary aquifers, what is true for local aquifers is even more so for aquifers underlying international boundaries which, in many cases, separate people of different nationalities, levels of development and political systems.

Governance of water resources

Governance is different from "management". When applied to water, management is typically oriented toward a single body of water or watercourse, and governance toward water as part of the resource base. Whereas management is often associated with the nitty-gritty of day-to-day operations, governance is concerned with how decision making is structured, how decisions are taken and who has influence over those decisions. In contrast to management, which emphasizes the results of decisions, governance emphasizes the processes by which decisions are made and put into effect. More fully stated by Nowlan and Bakker (2010, p. 7):

> Water governance is the range of political, organizational, and administrative processes through which communities articulate their interests, their input is absorbed, decisions are made and implemented, and decision makers are held accountable in the development and management of water resources and delivery of water services.

Governance is an old concept that has come back into common use over the past 15 years or so to recognize, among other things, the wider range of concerns that now impinge on governmental decision making, and the increasing role played by non-state actors in decision-making processes (Lautze *et al.* 2011).[1]

Several aspects of governance will be emphasized in this paper. First, as indicated by the quotation above, governance stresses the process rather than the results of decision making. Second, governance speaks to the entire range of this process, rather than the more narrow range of responsibilities entailed in management. Third, with its emphasis on process and with trends in many parts of the world toward greater public participation, governance involves a role for non-governmental organizations, community associations, the private sector and other stakeholders to complement traditional state-based actors. Fourth, and following from the three characteristics given above, governance requires a consideration of re-scaling decision-making processes, a topic that we discuss more fully below.

A number of forces are promoting a shift of water governance from governmental and managerial decision making towards more fluid and participatory approaches. The following list of such forces is adapted from Nowlan and Bakker (2010, pp. 7–8), and includes several that are particularly relevant to groundwater resources:

- Revised view of the role and mandate of government, in some cases associated with lack of confidence in public institutions;
- Greater awareness of the extent of expertise available outside government;
- Decreased government resources in comparison with the growing range of demands on water resources;

- Greater acknowledgement of the need to incorporate a range of values and perspectives in decisions about water;
- Greater acknowledgement of the need to integrate land-use planning and water resources management at a watershed scale.

The significance of these factors to the governance of transboundary aquifers is discussed below. But first we provide a few comments about the ways in which transboundary aquifers are distinct from intra-national aquifers and from surface water, and the consequences of those distinctions for governance.

Transboundary aquifers

We have known for a long time that there are many transboundary river basins –somewhere between 250 and 275 of them – and that they cover almost half of the globe's land surface (Wolf *et al.* 1999). (The number changes not because watersheds move but because borders do, as with the break-up of the Soviet Union.) Today, thanks to the work of the International Groundwater Resources Assessment Centre, a joint project of the United Nations Economic, Social and Cultural Organization (UNESCO) and the World Meteorological Organization, we can now say that there are at least as many transboundary aquifers as river basins.[2] Given the relative ignorance of groundwater as opposed to surface water in many parts of the world, one can be confident that the number of known transboundary aquifers will soon exceed the number of transboundary river basins.

International and intra-national transboundary aquifers

Blomquist and Ingram (2003) find a surprising number of analogies between intra-national and international aquifers. These analogies lead them to suggest that the process of building institutions for the two situations is also likely to exhibit many similarities. Governance of transboundary aquifers, they conclude, will be more difficult than, but not essentially different from, intra-national ones. However, in at least three ways, governance of international aquifers will be qualitatively different from governance of intra-national ones:

- First, by definition, some senior authority is available to mediate and, if necessary, resolve disputes over intra-national aquifers, at least for those that are large in volume or sizable in surface expression. Even in those areas of the world where groundwater extraction occurs without much government control, or even in defiance of it, national governments typically assert their authority once conflicts begin to occur or degradation of hydraulically linked surface water becomes an issue, or when border issues come into question (Norman and Bakker 2009). In contrast, senior authorities need to be created for international aquifers, and this in turn suggests the need to scale decision-making processes so they apply to the whole aquifer.
- Second, while the effectiveness of domestic regimes for managing aquifers is mixed, the problem can be addressed in principle by legislation and administrative procedure. In contrast, international aquifers must be governed at least in part by treaties, which are typically more difficult to negotiate and, if it proves necessary, to revise. More participatory modes of discourse and decision making could promote regimes that allow for more flexibility in the operation of such treaties;
- Third, although stakeholders in an intra-national aquifer may dispute the benefits of different ways of sharing the water, they are not likely to raise those issues to the

level of national security. In the case of international aquifers, national economic or military security is commonly invoked, which makes compromise in negotiations much more difficult, as is well illustrated in the Israeli–Palestinian case (Jägerskog 2003, Zeitoun 2008, Brooks and Trottier 2010).

Transboundary surface water and transboundary aquifers

Although it is generally accepted that most of the objectives, conventions and soft law that have been developed for transboundary surface waters will apply, with modest adaptations, to transboundary aquifers (Eckstein and Eckstein 2003, 2005), there remain important differences between shared surface water and shared groundwater resources.

The most obvious difference between surface water and groundwater is that the hydrometric and hydro-geological data required for sound decision making are significantly less available for the latter (Gaines *et al.* 2003). Some types of information, such as flow rates and directions, will be particularly important for transboundary aquifers. (For example, the Mountain Aquifer, most of which underlies the West Bank, is divided into three blocks, two of which emerge as springs in pre-1967 Israel, but one of which emerges as springs within Palestine.) The information gap for aquifers began to close in 2000 when the International Association of Hydrologists and UNESCO's International Hydrological Programme jointly launched the Internationally Shared Aquifer Resource Management (ISARM) Programme as a cooperative effort to inventory transboundary aquifers and to study their management (Puri and Aureli 2005). Given that drilling is expensive and unlikely to be possible in many areas, it would be helpful if ISARM could propose minimal types, quantities and qualities of data needed to permit adequate if not ideal conditions for decision making, with priority given to decisions involving transboundary aquifers. Special attention might also be given to water quality changes inasmuch as increases in pollution have been one of the drivers toward development of institutions to deal with transboundary surface water (Kliot *et al.* 2001).

Problems with lack of data have a further dimension in that, even with better information, agreements about surface water can literally be seen to be working (or not working), whereas those for groundwater are much less evident. The impact of excessive withdrawals, for example, can initially be largely invisible. (Indeed, for a time, drilling deeper wells or installing more powerful pumps can mask depletion.) In order to mitigate the climate of distrust that attends such invisibility, it is reasonable to conclude that any process that promotes transparency and sharing of data among a wide range of stakeholders will be conducive to successful governance of transboundary aquifers.

A second distinction between aquifers and surface waters with implications for governance is found in the greater vulnerability of groundwater to pollution and the difficulty of restoring aquifers once they have been polluted. The global record of groundwater pollution, ranging from influx of saline water to contamination by pathogens and persistent toxic substances, is well known and indicates that, if human use of these shared resources is to be sustainable, greater attention will have to be paid to the principle of "no significant harm", something that is best achieved through processes of governance, as discussed further below.

A third distinction is the significantly greater potential that an aquifer will be treated as an open-access resource. Typically, the added cost to drill deeper or to drill another well to obtain a greater share of the water is relatively small, but, if the same opportunity is available to everyone, the eventual, cumulative effect of competitive drilling is economic depletion of the resource at great cost to society at large. Though competitive exploitation can threaten any aquifer, the potential threat increases with the number of state borders (and thus the number of administrative agencies) overlying the aquifer. The Disi Aquifer, which is a large fossil aquifer underlying the border between Jordan and Saudi Arabia, was (and may still be) a prime example of the waste that can result from uncontrolled drilling on opposite sides of an international border (Long 2003, Ferragina and Greco 2008).

In order to avoid what has come to be called "the tragedy of the commons", some sort of common property regime or "unitization" agreement will have to be established and fostered for groundwater resources as they have been for other open-access resources (Ostrom 1994, Agrawal 2003, Weaver 2011). Such regimes will require that that resource users and managers develop common rules for exploitation that apply at the scale of the aquifer, something that is bound to be more difficult across an international border. In a different context, Bakker (2007, p. 447) has advocated recognizing and empowering alternative concepts of property rights, most notably in the form of what she calls "community economies of water"). The principles of governance that we are advocating in this paper would help provide a basis for establishing such regimes.

Re-scaling the governance of transboundary aquifers

As Nowlan and Bakker (2010, p. 8) point out: "Water governance has undergone a rescaling in the last decade. Frequently – but not always – this rescaling includes a shift towards watersheds. Many experts agree that shared water governance is best practised at the watershed level." Nowlan and Bakker nevertheless identify the many challenges involved in this approach while others have questioned its merits and highlighted the difficulties of implementing decision making at the watershed scale. (See Cohen and Davidson [2011] for a summary and advancement of this critique.) Notwithstanding the difficulties associated with surface water governance at the watershed scale, the ideal of matching governance processes to natural units has merit for aquifers, which differ from watersheds in several important respects: the physical properties of aquifers make them relatively discrete units that may or may not correspond to the physical contours of surface watersheds. They are not nested or subject to inter-basin diversion, and might thus be considered more natural units than watersheds. Moreover, their physical properties mean that those who share the use of an aquifer are more or less equally vulnerable to threats of drawdown and contamination – in effect a corollary of Salman's demonstration that downstream riparians can harm upstream riparians (Salman 2010). Aquifers may thus be appropriately regarded as common property resources, a status that implies the need for cooperative management at the scale of the aquifer itself. The rationale for governing at the scale of the aquifer itself is, if anything, even stronger in the case of transboundary aquifers.

Examples of emerging governance structures for managing transboundary aquifers seem to fit this rationale. For example, an international task force was developed for monitoring the Abbotsford-Sumas aquifer, which is shared by British Columbia and Washington State, and developing recommendations for shared management and pollution reduction (Norman and Melious 2004). The task force, which consists of representatives from different levels of government on both sides of the border, "addresses technical, scientific, and political issues related to the transboundary aquifer" and "coordinates local conferences and scientific exchange in order to foster heightened communication between key scientific, political and community players involved in the [shared aquifer]" (Norman and Melious 2004, pp. 105–106).

The Mountain Aquifer offers another example from a different and more highly contested part of the world. This limestone aquifer underlies the ridge on which several of the most important cities of Israel and Palestine are located. It is composed of three blocks, which flow in different directions: northeastward from the West Bank into Israel; westward from the West Bank into Israel and eastward almost entirely inside the West Bank. What is probably the most intensive attempt to develop a joint management regime for any transboundary aquifer was undertaken by an Israeli-Palestinian team of researchers in the 1990s (Feitelson and Haddad 1998, 2000). Their proposal treats the aquifer as a unit and provides for gradual extension of shared governance with inputs from stakeholders on both sides of the border until an all-inclusive agreement is achieved.

Special attention needs to be devoted to aquifers containing "fossil water", that is, water that accumulated centuries or millennia ago, and that remains contained within layers of rock that prevent recharge, or at best limit it to a tiny volume compared with what is stored in the aquifer. Any withdrawal is effectively depleting a non-renewable resource. (The same condition will occur if surface water sources are so low or evaporate so quickly as to limit recharge of the aquifer.) Some analysts have suggested that governance of aquifers containing fossil water requires a different approach from managing surface water, which is often assumed to be fully renewable over relatively short time periods (Eckstein and Eckstein 2003, Eckstein 2004). Clearly, in the case of fossil aquifers or aquifers with limited recharge, the notion of rescaling decision-making processes to the aquifer itself is even more pertinent, and should therefore be considered an important aspect of the approach.

Developing objectives

The concept of governance as the *process* by which decisions are made suggests that, rather than defining institutional models or structures in advance, the main focus should be on facilitating steps by which the various stakeholders would participate in forging structures and rules appropriate to their particular circumstances. Regardless of whether the focus is surface water or ground water, it is critical that all co-basin states agree on "equitable and reasonable use" of water by all parties. As is well known, this is easier said than done, and almost every water convention in recent years has reflected different views on how to apply that phrase (Salman 2007, Jones 2009, Rahaman 2009a, 2009b). With specific respect to transboundary aquifers, the Bellagio Draft Treaty (Hayton and Utton 1989) describes eight factors, ranging from hydro-geological and meteorological considerations to comparative costs and alternative sources of supply, which must be considered in defining equitable and reasonable in any specific case. Even this level of generality is qualified by noting that additional factors may also be considered, and by indicating that the weight given to any factor has to be determined by its importance relative to the importance of other factors.

The five objectives that follow can be seen as a complementary approach for realizing the broad goal of equitable and reasonable use. In keeping with our emphasis on governance as representing a more participatory and less technical approach than has been typical until now, these five objectives reflect more the social dimensions of transboundary aquifers than their physical and technical characteristics.

Social and political equity

Equity in some form is essential to the negotiation and ultimate adoption of any agreement related to the governance of shared water resources (Wolf 1999, Wohlwend 2001, Rahaman 2009a, 2009b). Rather than a simple quantitative sharing of water resources, a more nuanced form of equity is required in order to take into account what is *politically* acceptable to each side (Brooks and Trottier 2010). With the obvious exception of the relatively small volume of water required for drinking, cooking and bathing, the value of water to human society is found in the services that water provides rather than in the water itself (Gleick 1998, Brooks 2005, Brooks *et al.* 2009). Because those services have alternatives, there is a wide range of options for negotiation.

Wolf (1999) reports that successful agreements for sharing water have generally proceeded as each side gradually recognizes the demands of the other side(s), rather than by establishing a priori principles or rights. His observation is consistent with the views of other analysts that negotiations over water are increasingly shifting focus from existing uses to future needs. Notably, Lautze and Giordano (2006) argue that most successful agreements favour a needs-based rather than a prior-use basis. They also find (2006, p. 110) that international water agreements that appeal to terms and principles of equity are relatively successful because:

> . . . equity agreements are much more likely to include the provisions generally considered important to a treaty's resilience – conflict resolution mechanisms, amendment mechanisms, provisions for exchange of hydrologic data, and consideration of water quality.

The point we stress here is the importance of this process of mutual recognition, a process that requires engagement of and communication among a wide variety of actors and stakeholders – governmental and non-governmental – on both sides of a border. Fostering such a process differs from traditional inter-governmental negotiation in that it involves a wider range of participants.

In the search for equity, a dilemma that often arises is the need to balance the principle of equitable and reasonable use with a second, parallel principle that activities in one of the co-aquifer states not cause significant harm to the value of the resource for the other state(s) involved. Salman (2007) devotes a significant part of his article on rules and conventions for internationally shared water to the problem, and shows how differing formulations of the text of proposed agreements seem to give priority to one or the other principle, or, as in the case of the Berlin Rules, to indicate that they have equal standing. Wolf (1999, p. 7) provides a similar discussion, and states that the two principles "guarantee continued ambiguity in the principles of customary law".

Is there a way to minimize conflicts between two such important principles as equitable and reasonable use and avoidance of significant harm? We believe that there is in the case of transboundary aquifers. As emphasized above, among the most distinguishing features of aquifers is their sensitivity to pollution and the near impossibility of decontamination. We therefore suggest that for any aquifer and particularly for transboundary aquifers the principle of no significant harm must be met in order to realize the principle of equitable and reasonable use. In some cases, the flow of wastes into aquifers, whether inadvertent or deliberate, constitutes a clear form of unreasonable use. In other cases, harder pumping from lower portions of the aquifer can foreclose options of pumping from higher portions either by draining the aquifer or reducing pressure sufficiently to permit the influx of saline water. The case of the Disi Aquifer was mentioned above. In Israel, the Coastal Aquifer (not transboundary) is already so heavily affected by salts that 10% of the wells now carry more than 250 mg/l.[3] Along with growing levels of nitrates, this is a long-standing problem (Kanfi *et al.* 1983) that makes the water unsuitable for most irrigation and therefore represents an economic loss to the country. Also in Israel, the Mountain Aquifer (transboundary) is affected by poor control of sewage and solid waste in recharge areas, with effects on water quality in both Israel and Palestine (Tagar *et al.* 2005, Tagar and Qumsieh 2006). In short, in the absence of priority to the principle of no significant harm, equitable and reasonable use will, over time, become more and more difficult to achieve.

Economic efficiency

Textbook definitions of economic efficiency define it as a state in which all inputs are used in such a way as to maximize the value of outputs, as measured by monetary values or appropriate surrogates. While we acknowledge that efficiency is hardly a value-neutral concept (Trottier 2008), it takes on a particular importance in the context of internationally shared water resources. Each party to negotiations for a water-sharing regime needs to be confident that the other parties will use their share of the resource in what is perceived as an efficient rather than a wasteful manner. An example of an agreement based on this principle, albeit one involving surface water, is found in the Great Lakes Compact, which links the eight US states and the two Canadian provinces that border on the Great Lakes. The Compact has many clauses that require the parties to enact water conservation programmes and that can block further extraction unless the proposed use is economically efficient (Hall 2006).

The process by which mutual confidence in economic efficiency is established is therefore a critical aspect of the governance of shared aquifers, all the more so because it is difficult to follow economic rules strictly for water. A host of problems bedevil attempts to build agreements around economic efficiency – for example, the need to take into account water uses that fall outside the monetized economy. All too often, water is not valued until it is withdrawn from an aquifer, implying that the water in situ has no value. Thus, "beneficial use", as defined in much of western United States, only includes water that has been withdrawn. In recent years, analytical techniques have been developed to indicate the value of water in situ for aesthetics, for recreation and for ecological services to the stage where they can suggest limitations on withdrawals (Garrido and Livingston 2003, Postel and Richter 2003, Millennium Ecosystem Assessment 2005). The need to develop comparable criteria for aquifers and to apply these to discussions of economic efficiency presents an important challenge for governance.

With all of these issues precluding anything more than general agreement to be economically efficient, it should be no surprise that some concept of fairness typically trumps calculations of efficiency in international negotiations over water resources (Ingram *et al.* 1995, Syme *et al.* 1999, Wolf 1999, 2000, Blomquist and Ingram 2003). This openness can be taken as a starting point for developing flexible and mutually agreed approaches to what is adequately efficient within the framework of water governance.

Ecological sustainability

While ecological sustainability is obviously a critical objective of aquifer governance (Shah *et al.* 2000), the definition of "sustainability" is neither obvious nor objective (Molle 2009, Linton 2010). Budds (2009, p. 418), for example, has examined the role of "scientific knowledge" in state assessment of groundwater resources in La Ligua Basin of Chile to reveal "the subjective ways in which the assessment was interpreted and used by the state water resources agency to underpin water allocation decisions in accordance with its own interests . . ." Feitelson (2005, p. 413) shows why, even in Israel, where the knowledge of aquifer hydro-geology is high, "the political and economic incentives faced by decision-makers lead them to choose an unsustainable set of policies". The development of working criteria for sustainability of aquifers is everywhere difficult, and it is likely to be even more so for transboundary aquifers. Once again, the process by which these criteria are established presents a challenge as well as an opportunity for the governance of these resources.

Priority to demand management

There is no way that local or global water resources can satisfy all the demands that a modern society can put on them. Unfortunately, a supply mentality still pervades among many government bodies responsible for managing water (Furlong with Bakker 2007, Molle *et al.* 2009). As long as increasing water supply dominates thinking in water agencies, it tends to work against longer term economic, environmental, and likely even social sustainability (Gleick 2000, Gleick *et al.* 2003, Gleick *et al.* 2005, Brooks *et al.* 2009).

Successful management of shared aquifers will not be possible unless all parties emphasize demand management in order to help establish a climate of mutual trust that the shared resource is being valued and exploited equitably. The appropriate process of governance would make clear that, when a request is expressed for more water, attention must first go to determining whether the proposed new use of water can be satisfied from existing supplies or by policy and programme options that reduce the quality of water required for the end use, or that shift the timing of use to off-peak periods (Brooks 2006). In the context of internationally shared aquifers, such determinations involve a combination of technical and political negotiation.

Implementability

A final principle of successful governance of transboundary aquifers is to ensure that all interested parties have the institutional, social and financial means to participate in governance processes and translate agreements and rules into practice. Notably, institutions established to articulate interests, make decisions and oversee agreements at the scale of the aquifer itself would need to be given adequate authority and resources to function effectively, as would the various stakeholders who share an interest in the governance of the resource. Admittedly, this presents a very large challenge. Norman and Bakker (2009, p. 99) have shown that with respect to water governance along the Canada–United States border, "although a significant increase in local water governance activities has occurred since the 1980s, this has not resulted in a significant increase in decision-making power at the local scale . . ." And Fisher and Cook (2010, p. 467) point out while coordinating institutions are common for transboundary water bodies, "participating countries in general have not ceded any useful authority to the institutions they have created". Clearly, such institutions established for governance of transboundary aquifers will require adequate authority.

The realization of more participatory modes of decision making, which we have argued is central to the successful governance of transboundary aquifers, deserves particular attention. True, nominally participatory approaches to water governance do not necessarily empower local actors or even ensure equal representation among various stakeholders (for example, Cleaver [1999], Trottier [2008], Norman and Bakker [2009]). As Cleaver (1999, p. 609) has argued, the focus on participatory processes needs to be broadened: "A more dynamic vision is needed of 'community' and 'institutions' that incorporates social networks and recognizes dispersed and contingent power relations, and the exclusionary as well as inclusionary nature of participation."

While it is beyond the scope of this paper to elaborate such a vision, we suggest that attending more carefully to existing water-governance processes is one place to start. For example, the Oslo Agreements created the Palestinian Water Authority as a regulating body entrusted with implementing those parts of the agreement concerning water (Trottier 1999, World Bank 2009). However, about 70% of Palestinian water has long been managed by local communities or by associations of farmers, most often according to communal property regimes. These processes were made invisible by the new, apparently more "progressive" arrangements for water governance. Few Palestinians, moreover, even knew about this component of the agreement, and they continued to observe the practices of the grassroots institutions they perceived as legitimate.

One way of promoting successful participatory approaches to water governance is therefore to attend more carefully to sub-national, local and customary actors and practices when it comes to establishing rules for use of the resource (Trottier 1999, Wohlwend 2001, Brooks and Trottier 2010). Appropriate forms of governance would give local and customary forms of water management stronger standing – in effect, grant them "soft" or informal water rights, not unlike the "alternative concepts of property rights, most frequently some form of the 'commons' " as advocated by Bakker (2007, p. 441).

Conclusions

Governance focuses on processes for establishing means to manage transboundary aquifers rather than accepting an a priori desired final structure. Feitelson (2003), among others, argues that success with relatively simple agreements can lead to a positive cycle that then permits parties to consider more difficult ones. Some agreements, as with that on the Nubian Sandstone Aquifer, are limited to emergency measures, prior notification and information sharing. Sadoff and Grey (2005) are clear that, even if negotiations never get beyond these goals, they would still represent a significant gain. However, emphasis on process remains critical. Notably, instead of trying to derive specific criteria to accompany the principle of equitable and reasonable use, it will likely be more productive to look at actual agreements and infer from them the definitions of equitable and reasonable that have been applied (ideally, applied over time) to see how well they have worked and then identify from them possible models for application.

In this paper, we have argued that the growing need to manage transboundary aquifers in an equitable manner is best addressed through such processes. We have considered how international aquifers themselves constitute a scale at which interest-articulation and decision-making processes should take place in conjunction with other scales of governance. Good governance, as we have argued, should allow a growing role for non-state actors as participants in the decision-making process. We have therefore paid considerable attention to how participatory processes of decision making –especially at the scale of the aquifer – articulate with several objectives that are critical to the promotion of sustainable use of international groundwater resources. Along the way, we have urged that the trend to emphasize future needs (rather than existing uses) in negotiations about transboundary aquifers should be encouraged, and that, where a conflict exists, avoidance of significant harm should be met in order to realize equitable and reasonable use for governance of transboundary aquifers.

Finally, though by no means exclusive to aquifers or to international negotiations, it is critical to keep in mind Espelend's (1998) point that governance is not only a process, but a highly *political* process. After showing how political considerations affected water resource assessments by both Israeli and Palestinian scientists, Alatout (2000, p. 76) commented that: "The technical language of water potential cannot but be embedded in political meaning." Improving the governance of transboundary aquifers requires that the political meaning – indeed, political and social meanings – of water be respected by all parties.

Notes

1. For example, almost every one of the nine cases in the September 2010 special issue of *Water International* (35: 5) on Water, Food and Poverty in River Basins emphasizes *poor* governance as a key source of basin-wide problems.
2. See: http://www.igrac.net/dynamics/modules/SFIL0100/view.php?fil_Id=121.
3. See: http://hydroweb.com/israel-groundwater.html [Accessed 24 May 2011].

References

Agrawal, A., 2003. Sustainable governance of common-pool resources: context, methods and politics. *Annual Review of Anthropology*, 32, 243–262.

Alatout, S., 2000. Water balances in Palestine: numbers and political culture in the Middle East. *In*: D.B. Brooks and O. Mehmet, eds. *Water balances in the Eastern Mediterranean*. Ottawa: International Development Research Centre, 59–84.

Bakker, K., 2007. The "commons" versus the "commodity": alter-globalization, anti-privatization and the human right to water in the global south. *Antipode*, 39 (3), 430–455.

Blomquist, W. and Ingram, H.M., 2003. Boundaries seen and unseen: resolving transboundary groundwater problems. *Water International*, 28 (2), 162–169.

Brooks, D.B., 2005. Beyond greater efficiency: the concept of water soft paths. *Canadian Water Resources Journal*, 30 (1), 83–92.

Brooks, D.B., 2006. An operational definition of water demand management. *International Journal of Water Resources Development*, 22 (4), 521–528.

Brooks, D.B. and Trottier, J., 2010. Confronting water in an Israeli-Palestinian peace agreement. *Journal of Hydrology*, 382 (1–4), 103–114.

Brooks, D.B., Brandes, O.M., and Gurman, S., eds, 2009. *Making the most of the water we have: the soft path approach to water management*. London: Earthscan.

Budds, J., 2009. Contested H_2O: science, policy and politics in water resources management in Chile. *Geoforum*, 40 (3), 418–430.

Burke, J.J. and Moench, M.H., 2000. *Groundwater and society: resources, tensions and opportunities*. New York: United Nations Department of Economic and Social Affairs.

Burke, J.J., Sauveplane, C., and Moench, M., 1999. Groundwater management and socio-economic responses. *Natural Resources Forum*, 23 (4), 303–313.

Cleaver, F., 1999. Paradoxes of participation: questioning participatory approaches to development. *Journal of International Development*, 11 (4), 597–612.

Cohen, A. and Davidson, S., 2011. The watershed approach: challenges, antecedents, and the transition from technical tool to governance unit. *Water Alternatives*, 4 (1), 1–14.

Eckstein, G.E., 2004. Protecting a hidden treasure: the U.N. International Law Commission and the international law of transboundary ground water resources. *Sustainable Development Law & Policy*, 5 (1), 5–12.

Eckstein, G.E. and Eckstein, Y., 2003. A hydrogeological approach to transboundary ground water resources and international law. *American University International Law Review*, 19 (2), 201–258.

Eckstein, Y. and Eckstein, G.E., 2005. Transboundary aquifers: conceptual models for development of international law. *Ground Water*, 43 (5), 679–690.

Espeland, W.N., 1998. *The struggle for water, politics, rationality and identity in the American Southwest*. Chicago: University of Chicago Press.

Feitelson, E., 2003. When and how would shared aquifers be managed? *Water International*, 28 (2), 145–153.

Feitelson, E., 2005. Political economy of groundwater exploitation: the Israeli case. *International Journal of Water Resources Development*, 21 (3), 413–423.

Feitelson, E. and Haddad, M., 1998. *Identification of joint management structures for shared aquifers: a cooperative Palestinian-Israeli effort*. World Bank Technical Paper 415. Washington, DC: World Bank.

Feitelson, E. and Haddad, M., 2000. *Management of shared groundwater resources: the Israeli-Palestinian case with an international perspective*. Ottawa: IDRC Books; and Amsterdam: Kluwer Academic.

Ferragina, E. and Greco, F., 2008. The Disi project: an internal/external analysis. *Water International*, 33 (4), 451–463.

Fisher, M. and Cook, S., 2010. *Introduction* (to part one of special issue on water, food and poverty in river basins). *Water International*, 35 (5), 465–471.

Furlong, K., with Bakker, K., 2007. *Water governance in transition: utility restructuring and water efficiency in Ontario*. Vancouver: Program on Water Governance, University of British Columbia.

Gaines, L., Feitelson, E., and Wolf, A., 2003. Transboundary aquifers. *Water International*, 28 (2), 143–144.

Garrido, A. and Livingston, M.L., 2003. Economic and financial perspectives on intensive groundwater use. *In*: R. Llaman and E. Custudio, eds. *Intensive use of groundwater: challenges and opportunities*. Rotterdam: Balkema, 207–225.

Gleick, P.H., 1998. The changing water paradigm: a look at twenty-first century water resources development. *Water International*, 25 (1), 127–138.

Gleick, P.H., 2000. The human right to water. *In*: P.H. Gleick, ed. *The world's water 2000-2001: the biennial report on freshwater resources*. Washington, DC: Island Press, 1–17.

Gleick, P.H., Cooley, H., and Groves, D., 2005. *California water 2030: an efficient future*. Oakland, CA: Pacific Institute for Studies in Development, Environment and Security.

Gleick, P.H., *et al.*, 2003. *Waste not, want not: the potential for urban water conservation in California*. Oakland, CA: Pacific Institute for Studies in Development, Environment and Security.

Hall, N.D., 2006. Toward a new horizontal federalism: interstate water management in the Great Lakes Region. *Colorado Law Review*, 77 (2), 405–456.

Hayton, R.D. and Utton, A.E., 1989. Transboundary groundwaters: the Bellagio draft treaty. *Natural Resources Journal*, 29 (3), 663–722.

Ingram H.M., Laney, N.K., and Gillilam, D.M., 1995. *Divided waters: bridging the U.S.-Mexico border*. Tucson: University of Arizona Press.

Jägerskog, A., 2003. *Why states cooperate over shared water: the water negotiations in the Jordan River Basin*. Studies in Arts and Sciences 281. Linköping: Linköping University.

Jones, P., 2009. *The application of equitable and reasonable utilisation to transboundary water resources disputes: lessons from international practice*. PhD thesis. University of Dundee.

Kanfi, Y., Ronen, D., and Margaritz, M., 1983. Nitrate trends in the Coastal Plain Aquifer of Israel. *Journal of Hydrology*, 66 (1–4), 331–341.

Kliot, N., Shmueli, D., and Shamir, U., 2001. Development of institutional frameworks for for management of transboundary water resource. *International Journal of Global Environmental Issues*, 1 (3–4), 306–328.

Lautze, J. and Giordano, M., 2006. Equity in transboundary water law: valuable paradigm or merely semantics. *Colorado Journal of International Environmental Law and Policy*, 17 (1), 89–121.

Lautze, J., *et al.*, 2011. Putting the cart before the horse: water governance and IWRM. *Natural Resources Forum*, 35 (1), 1–8.

Linton, J., 2010. *What is water? The history of a modern abstraction*. Vancouver: UBC Press.

Long, B.R., 2003. *Middle East water conflict: the battle over the Al-Disi Aquifer* [online]. The Inventory of Conflict and Environment Case Studies. Washington, DC: American University. Available from: http://www1.american.edu/TED/ice/aquifer.htm [Accessed 2 January 2011].

Millennium Ecosystem Assessment, 2005. Freshwater ecosystem services. *In*: *Ecosystems and human well-being: current state and trends*. Washington, D.C: World Resources Institute. Volume 1, 213–255.

Molle, F., 2009. Water, politics and river basin governance: repoliticizing approaches to river basin management. *Water International*, 34 (1), 62–70.

Molle, F., Mollinga, P.P., and Wester, P., 2009. Hydraulic bureaucracies and the hydraulic mission: flows of water, flows of power. *Water Alternatives*, 2 (3), 328–349.

Norman, E.S. and Bakker, K. 2004. *Transboundary groundwater governance: an annotated bibliography* [online]. Vancouver: Program on Water Governance. Available from: http://www. water-governance.ca/PDF/h20_governance_annotated_bibliography.Norman.Baker.pdf [Accessed 13 December 2010].

Norman, E.S. and Bakker, K., 2009. Transgressing scales: water governance across the Canada-U.S. borderland. *Annals of the Association of American Geographers*, 99 (1), 99–117.

Norman, E.S. and Melious, J.O., 2004. Transboundary environmental management: a study of the Abbotsford- Sumas aquifer in British Columbia and western Washington. *Journal of Borderlands Studies*, 19 (2), 101–119.

Nowlan, L. and Bakker, K. 2010. *Practising shared water governance in Canada: a primer* [online]. Vancouver: University of British Columbia Program on Water Governance. Available from: http://www.watergovernance.ca/wp-content/uploads/2010/08/PractisingSharedWaterGovernancePrimer_final1.pdf [Accessed 13 December 2010].

Ostrom, E., 1994. *Neither market nor state: governance of common pool resources in the 21st century*. Washington, DC: International Food Policy Research Institute.

Postel, S. and Richter, B., 2003. *Rivers for life: managing water for people and nature*. Washington, DC: Island Press.

Puri, S. and Aureli, A., 2005. Transboundary aquifers: a global program to assess evaluate, and develop policy. *Ground Water*, 43 (5), 661–668.

Rahaman, M.M., 2009a. Principles of international water law: creating effective transboundary water resources management. *International Journal of Sustainable Society*, 1 (3), 207–223.

Rahaman, M.M., 2009b. Principles of transboundary water resources management and Ganges treaties: an analysis. *International Journal of Water Resources Development*, 25 (1), 159–173.

Sadoff, C.W. and Grey, D., 2005. Cooperation on international rivers: a continuum for securing and sharing benefits. *Water International*, 30 (4), 420–427.

Salman, S.M.A., 2007. The Helsinki Rules, the UN Watercourses Convention and the Berlin Rules: perspectives on international water law. *International Journal of Water Resources Development*, 23 (4), 625–640.

Salman, S.M.A., 2010. Downstream riparians can also harm upstream riparians: the concept of foreclosure of future uses. *Water International*, 35 (4), 350–364.

Shah, T., *et al.*, 2000. *The global groundwater situation: overview of opportunities and challenges*. Colombo: International Water Management Institute.

Syme, G.J., Nancarrow, B.E., and McCreddin, J.A., 1999. Defining the components of fairness in the allocation of water to environmental and human uses. *Journal of Environmental Management*, 57 (1), 51–70.

Tagar, Z., Keinan, T., and Qumsieh, V., 2005. *Sewage pollution of the mountain aquifer: finding solutions*. Amman, Bethlehem and Tel Aviv: Friends of the Earth Middle East.

Tagar, Z. and Qumsieh, V., 2006. *A seeping time bomb: pollution of the mountain aquifer by solid waste*. Amman, Bethlehem and Tel Aviv: Friends of the Earth Middle East.

Trottier, J., 1999. *Hydropolitics in the West Bank and Gaza Strip*. Jerusalem: Palestinian Academic Society for the Study of International Affairs.

Trottier, J., 2008. *Water crisis: political construction or physical reality? Contemporary Politics*, 14 (2), 197–214.

Weaver, J.L., 2011. *Unitization of oil and gas fields in Texas: a study of legislative, administrative and judicial policies*. Washington, DC: Resources for the Future.

Wolf, A.T., 1999. Criteria for equitable allocations: the heart of international water conflict. *Natural Resources Forum*, 23 (1), 3–30.

Wolf, A.T., 2000. From rights to needs: water allocations in international treaties. *In*: E. Feitelson and M. Haddad, eds. *Management of shared groundwater resources: the Israeli-Palestinian case with an international perspective*. Ottawa: IDRC Books; and Amsterdam: Kluwer Academic, 27–59.

Wolf, A.T., *et al.*, 1999. International river basins of the world. *International Journal of Water Resources Development*, 15 (4), 387–427.

Wohlwend, B.J., 2001. *Equitable utilization and the allocation of water rights to shared water resources* [online]. Unpublished MS. Available from: http://www.bjwconsult.com/EQUITABLE.pdf [Accessed 2 January 2011].

World Bank, 2009. *Assessment of restrictions on Palestinian water sector development*. Sector Note for West Bank and Gaza. Washington, DC: World Bank.

Zeitoun, M., 2008. *Power and water in the Middle East*. London: I.B. Tauris.

Interpreting the unknown: uncertainty and the management of transboundary groundwater

Anita Milman and Isha Ray

This paper shows how uncertainty undermines collaborative transboundary groundwater management. Focusing on the Santa Cruz Aquifer, spanning the United States–Mexico border between Arizona and Sonora, the authors describe the uncertainties within the aquifer using interviews and hydrologic studies. We discuss how data requirements and ambiguous interpretations exacerbate these uncertainties, and explain how each country's water-management culture combines with this uncertainty to create contrasting views on groundwater availability and abstraction impacts. As a result, water managers in both countries predict different impacts from pumping and recharge, and each uses that information discursively to support unilateral policies rather than to promote collaborative management.

Introduction

Many experts on cooperation over transboundary waters promote norms such as "equitable and reasonable use", "no significant harm" and "optimal basin development" (United Nations 1997, Wolf 2007); for a countering view see Wegerich and Olson (2010). To be useful, these norms require shared knowledge of hydrologic processes and shared understandings of "equity" and "harm". Our research in the transboundary Santa Cruz Aquifer, located along the United States–Mexico border, makes evident the difficulty of applying these norms to shared groundwater. Groundwater systems are characterized by inadequate information (epistemic uncertainty). Epistemic uncertainty can lead to framing uncertainty, or ambiguities in how a situation is understood and represented. We argue that poor data and clashing frames make it unclear how groundwater cooperation is to be achieved, if needed, or even how to determine the value (if any) to cooperative management.

Our research was conducted from 2005–2008 and included more than 40 key informant interviews, a detailed analysis of hydrogeologic data and groundwater simulation modelling. We begin with a discussion of the complexities of groundwater flow and their portrayal in the literature. We then describe the study region, explaining each country's perceptions of water availability and its conceptualization of the hydrology of the aquifer. We then show how inadequate data and institutional factors produce a poorly understood aquifer and explain how framing allows each country to hold different perspectives on the same aquifer. Each interprets information on the hydrologic processes via the lens of its own water-management paradigms and moves forward with its individual agenda. Lastly, we discuss the implications of our findings to the Santa Cruz aquifer and to transboundary aquifers more generally.

Uncertainty is inherent in transboundary groundwater

Uncertainty is characterized by the manner in which it can be reduced. Aleatory uncertainty refers to that which is stochastic in nature, and therefore cannot be reduced, although it may be described via probability-distribution functions. Epistemic uncertainty is due to insufficient knowledge of a system, and could be reduced, although the parties concerned may be unable or unwilling to do so (Oberkampf *et al.* 2004, de Rocquigney *et al.* 2008). In the international-relations literature, epistemic uncertainty is often called model uncertainty; this is contrasted with strategic uncertainty, referring to uncertainty on account of asymmetric information held by the bargaining parties (Iida 1993). Brugnach *et al.* (2008) add to this typology the concept of framing ambiguity, which refers to uncertainty arising from multiple concurrent framings or interpretations of a particular phenomenon.

Although all water-resource systems are characterized by aleatory uncertainty stemming from natural variability in climate, uncertainty in groundwater is also strongly epistemic. Groundwater systems are complex, open systems; therefore they are underdetermined (Oreskes *et al.* 1994). Data collection is expensive and requires technical expertise. Key features of the aquifer that impact flows, such as fractures, impermeable regions and high or low conductivity lenses, complicate measurement, as do stresses on the system that, by shifting gradients, may lead to induced recharge and capture of streamflow (Bredehoeft 2002). Transboundary aquifers are particularly difficult to characterize because information must cover the extent of the aquifer across multiple countries and data on piezometric levels, and flows taken from each side of the border should cover a similar timeframe and sampling frequency (see for example, Anderson and Woessner [1992]). But water agencies across a border operate independently, so such coordinated data usually do not exist. As a result, co-aquiferians do not have full (or even the same) understanding of how groundwater flows through the shared aquifer: they may hold different conceptual models of the aquifer, leading to different interpretations of water availability, and the impacts of groundwater use, recharge and protection activities. In the Santa Cruz, this epistemic uncertainty is amplified by multiple frames, as sparse data are interpreted via the lens of water-management paradigms.

The high degree of uncertainty associated with groundwater is well understood by hydrologists (Tsur and Zemel 1995, Beven 2000). Multiple working hypotheses are, in fact, fundamental to the science of hydrogeology (Chamberlin 1897, Bredehoeft 2005). Yet there is little explicit analysis of the role of uncertainty in studies of transboundary water management. Models of transboundary water management presume that countries share a common physical frame (hydrologic model), through which they know the quantity of water physically available as well as the impact possible water-management strategies will have on hydrologic processes. This assumption of shared and agreed-upon knowledge is especially common in economic approaches to transboundary waters, such as game theoretic (Frisvold and Caswell 2000, Dombrowsky 2007), optimization (Küçükmehmetoğlu and Guldmann 2004, Whittington *et al.* 2005), and benefit-sharing analyses (Sadoff and Grey 2005, Phillips *et al.* 2006). With few exceptions (for example, Fischhendler [2004]), variability in flows is rarely considered, and uncertainty of hydrologic processes has yet to be acknowledged in these policy-oriented literatures. This is especially true for ex-ante uncertainty that determines whether or not a collaborative treaty is possible, or even needed (as opposed to ex-post uncertainty that relates to treaty effectiveness.[1]

The clearest recognition of the role of ex-ante uncertainty in collaboration is by Lemarquand (1976) who suggested that when the consequences of a transboundary water-management strategy are not fully understood, countries may be reluctant or unwilling to reach an agreement. Zeitoun and Warner (2006) also suggest that ambiguity can be manipulated during the process of constructing "knowledge", a key mechanism leading to hydro-hegemony. Our research in the Santa Cruz highlights how multiple uncertainties brings to the fore extra-scientific considerations, such as equity, sovereignty, prestige and cultural ideologies. These considerations are, of course, known to influence the water-management strategies that countries adopt (Frey 1993, Wolf 1997, Bernauer 2002), but have been analytically sidelined by the emphasis on benefit sharing and Pareto-optimal basin management. We argue that these value-laden factors are key to interpretations of what benefits are to be shared or optimized, and thus influence the possibilities for collaboration over shared groundwaters.

The Santa Cruz Aquifer

The Santa Cruz Aquifer is located along the United States–Mexico border, spanning parts of Arizona and Sonora. There is a strong history of cooperation between the two countries over their shared surface waters, but little precedent for management of groundwater in the 17 aquifers that traverse the border (Mumme 2005). An addendum to the 1944 treaty that serves as the basis for water relationships between the two countries implied that a comprehensive groundwater agreement would be forthcoming (IBWC 1973), yet such an agreement has not been drafted.

The study region is defined by the path of the Santa Cruz River and the aquifer connected to it, as depicted in Figure 1. On the US side of the border, the region encompasses the area defined as the Santa Cruz Active Management Area (SCAMA). On the Mexican side it encompasses the municipalities of Nogales and Santa Cruz, Sonora. The region, which is part of the Sonoran Desert, is semi-arid and drought-prone (Morehouse *et al.* 2000, SAGARPA 2004). Precipitation is between 280 and 440 mm per year (Liverman *et al.* 1997) and occurs over two seasons, the summer monsoons (August/September) and the winter rains (December through February).

The aquifer is made up of three hydrogeologic units: the Younger Alluvium (YA), also considered the alluvial floodplain aquifer; the Older Alluvium (OA), also known as the upper basin fill; and the Nogales Formation (NF), known as the lower basin fill. In several locations, outcrops of the Nogales Formation or bedrock constrictions create narrows, and pockets known as microbasins that act as mini-storage basins (Erwin 2007, geologist, pers. comm. 13 October 2007). The aquifer is characterized as unconfined or semi-confined and water-table levels fluctuate seasonally. There is a strong interaction between surface and groundwater, and the majority of water used in the region is withdrawn from wells located in the river valley.

Water availability and the impacts of use on water-table levels are primary concerns in the region. Up until the recent economic crisis (2007–2009), both sides of the border were experiencing rapid growth and that trend is likely to continue. While there is no immediate crisis, water agencies in both Arizona and Mexico recognize that the demands on the shared aquifer will intensify in the next few years. In Arizona, the priority is to meet increasing demands for housing and economic growth while protecting ecosystems and maintaining the aesthetic and cultural aspects of the community (ADWR 1997, Santa Cruz County 2004). Here water-management policies centre on preventing water-table

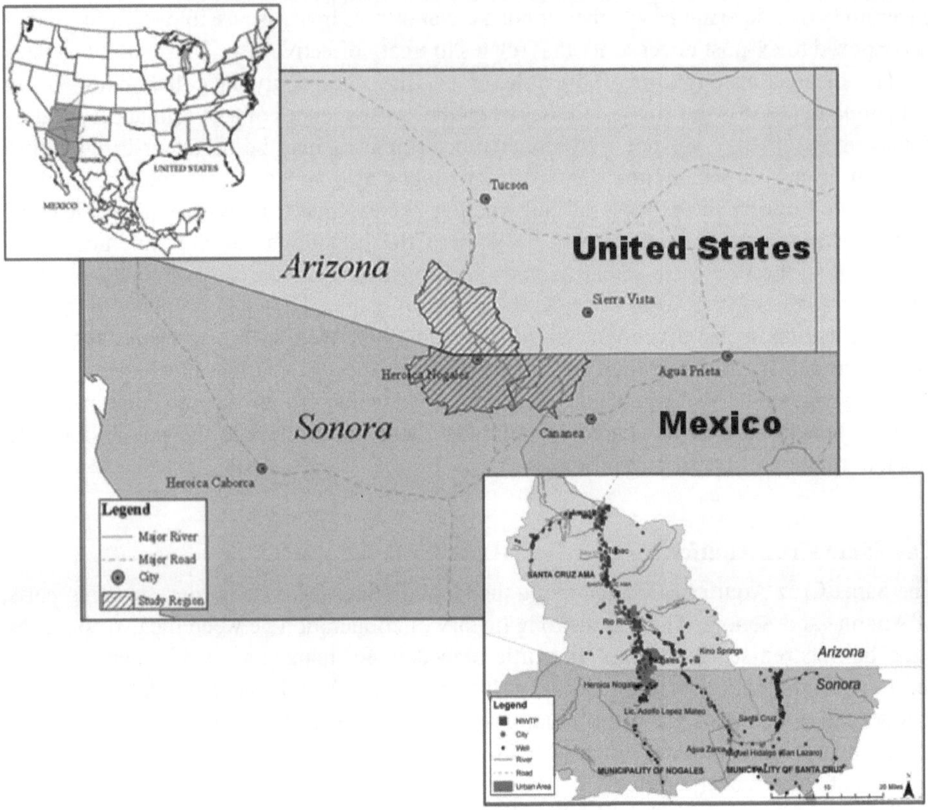

Figure 1. The Santa Cruz Aquifer: administrative boundaries and key features.
Source: Milman and Scott (2010, p. 530).

drawdown, maintaining safe yield and ensuring sufficient supply for growth. In Sonora, the priority is to improve and extend water-supply services; as of 2007, 13% of residents in the urban area of Nogales did not have access to piped water.[2] Sonora also wants to ensure that its burgeoning industrial sector continues to have its water needs met. Thus the focus is on transporting water from the aquifer to the population through infrastructure development. The potential forums for interaction over the aquifer include managing the timing, volume and location of groundwater abstractions, transfer of water across the border (likely from Arizona to Mexico) to mitigate the geographic impacts of pumping, and conducting artificial recharge of the aquifer.[3]

Views differ on water availability and impacts of use

From conversations with water managers on each side of the border and observation of water-planning meetings it is evident that perspectives on water availability and the impacts of pumping differ. Within Arizona, it is widely commented that groundwater pumping in Mexico adversely impacts both water availability and stream flow in Arizona. In 1995, an Arizona Department of Water Resources (ADWR) evaluation of future Mexico well-field expansion reported: "Increased pumpage of the Mexican well fields on the Santa Cruz River will have a detrimental effect on the ability of wells on the US side of the border to obtain water" (ADWR 1995a, p. 1). The same report suggested that increased pumping in Mexico could result in future supply shortages and damage to the riparian ecosystem, including habitat for several endangered species. Interviews[4] with large water-rights holders, developers and environmentalists indicate that they concur with this negative assessment. They expressed concern about the future, about the rapidly increasing demands for water within Mexico and the impact that meeting those demands could have on their water supply and ecosystems. As one interviewee describes the situation: "We've got to come to some sort of agreement with Mexico, the SCAMA cannot go on a lot longer with the dark cloud of what is going to happen hanging over our heads" (10 October 2007).

Water specialists in Arizona also acknowledge uncertainty in the conceptual model[5] of the aquifer. One interviewee, widely known to be an expert on the hydrogeology of the Santa Cruz, explained that the role of fractures in the bottom layer of the aquifer (the Nogales Formation) in conveying water was unknown, and could greatly influence estimates of underflow across the border. Moreover, although (observed) declines in the summer base flow at the Nogales stream gauge (Shamir *et al.* 2005) may be attributable to capture by Mexican pumping, they may also be caused by other changes, such as channelization of the river or increased riparian vegetation (ADWR, pers. comm. 28 April 2007). Despite the general belief in the negative impact of pumping in Mexico, the complexity of the aquifer is well recognized in Arizona.

Water managers in Mexico express a different perspective: they assert that additional water can be abstracted from the aquifer and this abstraction will have minimal impact on the United States. Although residents on the Sonora side of the border experience water shortages,[6] water managers explain this not as a result of insufficient water but rather of insufficient funding for the construction and repair of conveyance infrastructure (OOMAPAS[7], pers. comm., 14 July 2005, 21 and 22 September 2007). The view that there is additional water available for withdrawal is supported by a study of the aquifer by the Mexican National Water Commission (Comisión Nacional de Agua, hereafter CONAGUA), which determined that current withdraws are less than annual average recharge (CONAGUA, n.d.). Overall, Mexican officials believe that the cross-border effect of their water use, if any, is minimal. Interviewees stated they expect no significant impact on Arizona because the aquifer is shallow, has a high conductivity and the Santa Cruz River is (and remains) a gaining stream. These factors imply that there may be physical limits to the potential impact of pumping across the border, but they do not rule out negative cross-border impacts of increased groundwater use.

The differing perspectives each country holds regarding the availability of groundwater and the impacts of its use could stem, in part, from its water-management priorities and its national self-interests. At the same time, these divergent perspectives are a result of (and subsequently bolstered by) incomplete understandings of the hydrologic processes combined with ambiguity regarding the concept of safe or sustainable yield. As is discussed below, the determination of safe or sustainable yield has been a topic of much debate by hydrogeologists (Bredehoeft 1997, Zhou 2009). Consequently, safe or sustainable yield is itself subject to interpretation via the lens of each country's water values and self-interest.

Data sparseness allows for divergent perspectives

In the Santa Cruz, there is insufficient empirical data to characterize the hydrogeologic properties of the aquifer and to describe flow processes. The lack of data sharing across the border augments this uncertainty. Even were information sharing to be the norm, the frequency and timing of measurements (piezometric data, abstractions) on each side of the border are uncoordinated, so using that information would be challenging. As result of sparse data (and the subjectivity of its interpretation), the conceptual models of the aquifer that were developed independently by each country contain gaps and do not match.

In the conceptual model developed by Arizona, the hydraulic connections between the layers and vertical hydraulic conductivities are unknown, and flow patterns outside the younger alluvium have to be assumed (Erwin 2007, Nelson 2007). A paucity of measurements, missing or "ambiguous" (Erwin 2007, p. 84) drilling logs and insufficient testing mean they have no information on how water levels vary between the hydrogeologic layers of the aquifer, and limited knowledge of the thicknesses and extents of the three hydrogeologic units (Erwin 2007, Nelson 2007). Recharge and evapotranspiration values are uncertain on account of scarce and unreliable pumping data (Nelson 2007), and mountain-front recharge processes are similarly uncertain (Wilson and Guan 2004).

The conceptual model developed for the Mexican side of the border is fraught with even more gaps, as data is extremely limited; see Table 1 below for a summary of those gaps. There is little historical data on piezometric head, flux into and out of the aquifer, or tributaries and stream flow. The only data on water levels is from three well censuses (which were taken several years apart and during different seasons: November 1989, December 1997 and August 2002) and from measurements of water levels in seven of the wells operated by OOMAPAS in May and June 2005.[8] Measurements of flux between the river and the aquifer do not exist; the only stream gauge on the Mexican side of the border was damaged by flooding in 1974, repaired in 2000, and destroyed shortly after by vandals (Comisión Internacional de Límites y Aguas (CILA), pers. comm. 2 October 2007). Historical groundwater abstraction information is limited to estimates from the well census for 1997.[9] Data on well construction also does not exist (see Milman [2009, ch. 5]) and hydrogeologic data on aquifer properties and geometry is similarly sparse (SeismoControl 1995, SeismoControl 1996, Milman 2009).[10]

Not only are there gaps in the conceptual models developed within each side of the border, but those models also differ dramatically. Figure 2 shows the conceptualization of the aquifer at the international boundary developed by ADWR (Erwin 2007) for the US side of the border, and the model developed by a sub-contractor for CONAGUA for the Mexican side of the border. Both countries agree that the aquifer is made up of three hydrogeologic layers, yet the conceptualization in the Arizona model is of shallower, less conductive layers whereas the Mexican model depicts a deeper aquifer with greater conductivity. No hydrogeologic tests have been conducted at this point, so the conceptualization is based on interpolation of data by each country. In addition to differences in transmissivity,[11] the model developed for the Mexican side of the border estimates a recharge rate equivalent to 19% of annual average precipitation, while the model for Arizona estimates a recharge closer to 4% (SeismoControl 1996, Milman 2009).

In groundwater-simulation modelling, conductivity parameters are most frequently calibrated to account for uncertainties arising from heterogeneity in the soil and interpolation of aquifer layer thicknesses. Given equifinality,[12] it is not unusual that the models have differing depths and conductivities; what is unusual is in this case is the magnitude of these differences. The aquifer as conceptualized in the Mexican model is nearly twice the thickness of that in the ADWR model and contains a higher conductivity. The Mexican groundwater-simulation model also assumes the aquifer is in steady-state equilibrium, whereas the ADWR simulation model is a transient model. The conceptualization from the Mexican side suggests an aquifer through which a larger quantity of water can flow more easily, reflecting the perspective of water managers that water availability is not the problem.

Table 1. Epistemic uncertainty: gaps in the conceptual model of the Santa Cruz Aquifer for the Sonora side of the border.

Aquifer system property	Uncertainties in the conceptual model
Estimates of underflow at system boundaries (into and out of the aquifer)	Estimates only exist for the top layer of the aquifer at the northern model boundary (international border). Estimates for other layers of the aquifer at the northern model boundary cannot be developed due to a lack of data on the thickness of and groundwater gradient in those layers. No information exists regarding layer thicknesses or groundwater gradients for the southern model boundary.
Inter-connections between aquifer layers	Limited piezometric head observations exist from which to determine the presence (and direction) of vertical flow between the layers. Pumping tests do not extend to the older alluvium or the Nogales formation. Reports and models are incongruent in their characterization of the lower layers of the aquifer as unconfined or confined. Studies indicate possible fractures within the Nogales formation. The Santa Barbara well field was drilled on a fault, but this fault and other potential faults are not included in the conceptual model.
Streams and stream–aquifer interaction	Very few measurements of stream flow and stream stage exist within the study region making it difficult to estimate baseflow, stream flux and reactions to recharge events. No measurements of tributaries exist, nor is it clear if any are perennial.
System geometry	Cross-sections only exist for the western portion of the aquifer and do not include the boundaries of the region to be modelled. Microbasins are known to exist in the region yet may not be accurately depicted due to the limited cross-sectional data available.
Historical water levels and fluxes	Comparison of 1989 and 1997 water levels suggest the aquifer may be in steady state, but this analysis did not account for changes in precipitation/recharge or pumping during this time period. The 2002 water-level observations occurred in the middle of the summer monsoon season, and due to seasonal variation in the aquifer, comparing those with the 1989 or 1997 values would provide little information.
Historical abstraction rates	Data reports listing historical abstraction rates are inconsistent.
Recharge mechanisms and quantities	Studies of recharge in the arid US southwest suggest mountain-front recharge may be the dominant recharge mechanisms in the region. Limited groundwater-level measurements and a lack of a time series for individual wells make it difficult to determine the relationship between precipitation/recharge events and groundwater levels. Conceptual estimates of tributary recharge do not exist, nor is there sufficient data on tributaries and flow to develop such estimates. The lack of estimates for underflow into and out of the older alluvium and Nogales formation, combined with unknown historical pumping rates and tributary flows, makes it difficult to estimate recharge inversely though a water-balance calculation.

Note: See Milman (2009, chapter 5) for a complete analysis. Data from: Erwin (2007); SeismoControl (1995, 1996); Padilla (2005); CONAGUA (2009, nd); Shamir *et al.* (2005); and personal communications including raw data from CONAGUA, OOMAPAS, ADWR and local geologists.

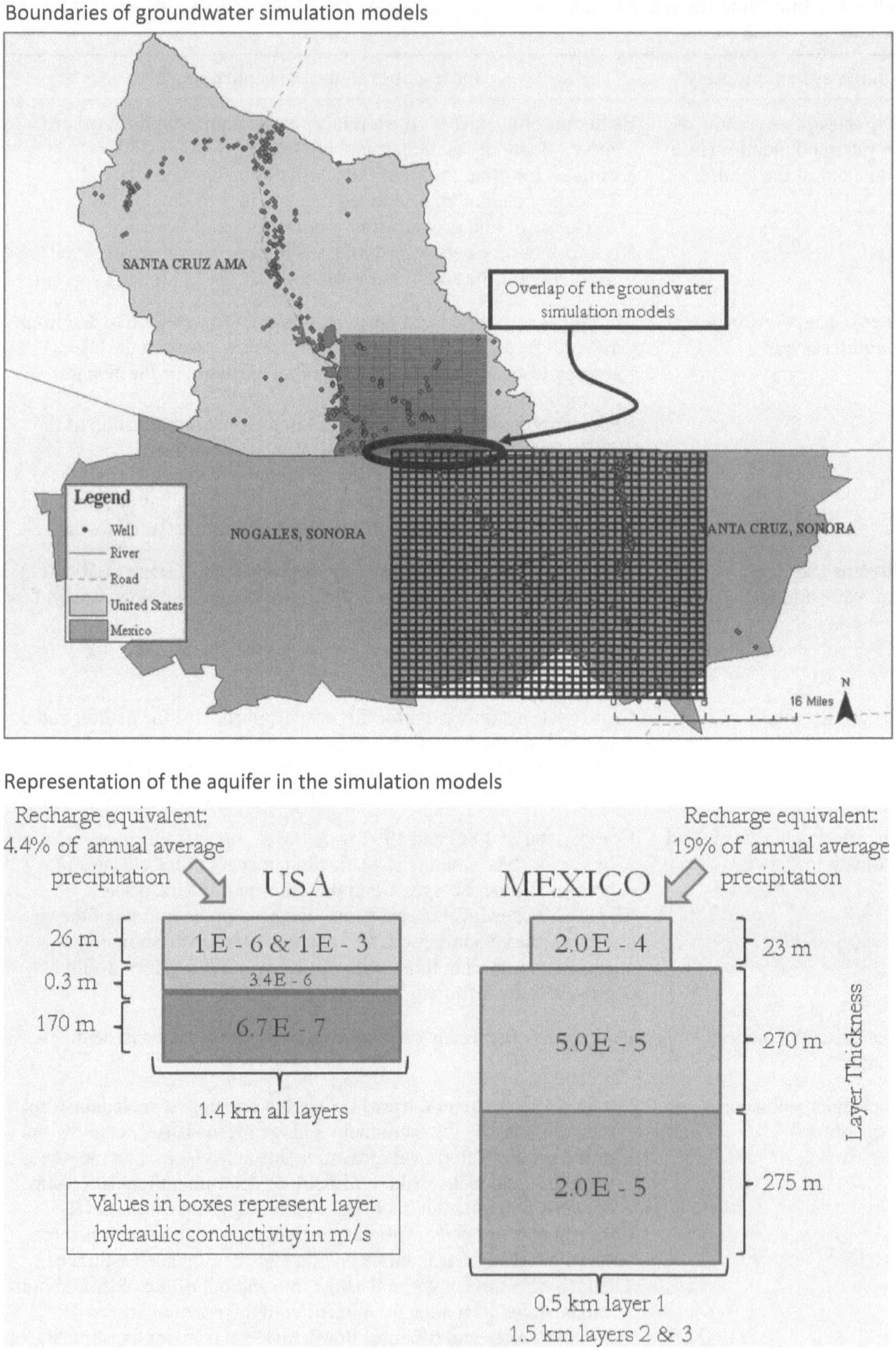

Figure 2. Extent and details of the groundwater simulation models developed for each side of the border.

Water management activities are significantly value-driven

Compounding the effects of divergent conceptual models, uncertainty also stems from ambiguity in definitions of "safe" yield. One common conception of safe yield is as a quantity determined by natural recharge; this definition is premised on the depletion of storage from any pumping greater than recharge. A second common perspective is that the amount of groundwater available depends not on natural recharge, but rather on capture (Bredehoeft 2002). This definition accounts for the fact that an aquifer contains multiple equilibrium states because changes in groundwater gradients induce water to flow into the aquifer. It also considers that stream depletion may be considered an acceptable way to increase water availability, as, in some environments, storage of stream water in the aquifer will reduce evaporative losses. Yet even pumping at rates less than (natural and induced) recharge can lead to declining water levels, depletion of stream flow and ecosystem deterioration (Zhou 2009). Consequently, the definition of "sustainable" (rather than "safe") yield considers the response of the aquifer to pumping and the effects of this response on the environment and society (Kalf and Woolley 2005, Zhou 2009).

In the Santa Cruz Aquifer, the water values held by each country influence the definition of "safe" yield it has adopted and consequently, the management activities it undertakes. Arizona's culture of water management centres on the use of mathematical models for decision making to protect growth and the environment, as well as existing water users. Priorities include protecting the vibrant riparian corridor, protecting the habitat of several endangered species,[13] preserving the character of the community, and requiring assured water supplies before further development is permitted. Thus in Arizona, safe yield includes the requirement that water tables cannot fall below a specified level, so that pumping costs are contained and (in theory, at least) the ecosystem is protected. As a result, the SCAMA determines water availability by considering both water balances and localized declines in water-table levels, defined as drawdown that exceeds one standard deviation of historically measured levels (ADWR 1995b, Corkhill 2006, ADWR 2007).

In Mexico, the emphasis of water management is on addressing unmet human needs (as the majority of the population does not have 24-hour water service) and prioritizing future growth and development. This prioritization was expressed by one official who said: "The main thing in Mexico is we still haven't developed our supplies, we have not yet gotten to worries about trees and the environment. We still need water for people" (meeting for the Transboundary Aquifer Assessment Program, 28 January 2008). In addition, the Mexican Constitution declares water to be the "patrimony of the Mexican people" (Farias 1993, Instituto Mexicano de Technología del Agva (IMTA), pers. comm. 14 June 2006), a sentiment regularly echoed by interviewees. The water located in Mexico, they argued, should be used to best benefit the Mexican population, and should not be relinquished to the United States. These values are reflected in the Mexican calculation of water availability, which is based on an average annual water-balance analysis and does not consider annual variation in water-table levels (CONAGUA, n.d.), as well as by the recommendation to capture for human use the water currently being "wasted" as evapotranspiration by riparian vegetation (SeismoControl 1996).

Both these operational definitions for yield base the amount of water available for withdrawal on an estimate of renewable supplies and neither is designed to lead to groundwater mining. Yet they result in divergent perspectives on availability. Moreover, they inform the management activities each country undertakes. In line with its values, ADWR is using groundwater simulation models it developed in 2007 to conduct stochastic simulations to improve their ability to regulate water level drawdown (ADWR, pers. comm., 30 October 2007). It is also coordinating with researchers from other institutions to investigate stream-aquifer interactions,[14] causes of tree die-off in the region[15] and the feasibility of conducting aquifer recharge and storage.[16] In addition, preliminary discussions are under way on comprehensive water-management planning, including formation of an entity with the legal authority to lease store or sell water across the border, to secure supplies and protect the environment (water lawyers, pers. comm. 17 and 18 October 2007, ADWR, pers. comm. October 2007).

Water-management activities in Mexico stem from its values that unmet human needs take priority. Efforts to improve water-supply services include the construction of phase two of the "Acuaferico" project,[17] which aims to upgrade and extend water-system services via the drilling of new wells, and construction of new pumping stations, water tanks and distribution lines. Smaller projects are also in progress, which will connect several neighbourhoods to the citywide piped-water network. Increases in water abstraction are being monitored and CONAGUA has been active in registering and regulating concessions of water, especially those providing water for delivery by tanker trucks. On the Mexican side of the border, the agencies are also working to install water meters, repair leaks, rehabilitate wells and educate the public regarding water conservation.

The different definitions of sustainable yield and the culture of water held by each country are reflected in their transboundary plans and practices. On the Mexican side of the border, the plan is to increase aquifer pumping. In contrast with Arizona's fears, no detrimental cross-border effects are expected. This difference arises in part because Mexico prioritizes addressing unmet needs, and in part because the conceptual model of the aquifer held by the Mexican water managers suggests a greater availability of water than the model used on the Arizona side. The difference is not unrecognized, as during an interview, an Arizona government official quoted a Mexican official as having said: "You [Arizona] want water for your fish, but here in Mexico we have people without water to drink. God will punish you" (2 June 2006). Whatever the accuracy of this recounting, the interpretation of it by the Arizona official suggests that the different culture of water held by each country is understood across the border. From the perspective of Mexico, an agreement that allows Arizona to improve water availability for fish, but that does not improve water availability for the Mexican population, would be unthinkable.[18]

The US Congress-funded Transboundary Aquifer Assessment Program (TAAP) has sparked some movement towards collaboration in the form of data collection and sharing. The Santa Cruz Aquifer is one of four priority aquifers listed in the TAAP (United States Senate 2006). The programme does not represent a formal agreement on transboundary groundwater management; rather it is a mechanism for funding joint investigations and has led the US and Mexico to reach an agreement on joint data collection. The hope is that the programme may be instrumental both in reducing framing uncertainties and in the formation of cross-border networks and social capital, hypothesized to be key components of cooperation (Young 1989, Blatter and Ingram 2001). The effectiveness of the TAAP has yet to be determined, as even three years after its approval it has progressed little due to funding constraints.[19]

Conclusions and implications

We have argued that the sparseness of data on the Santa Cruz Aquifer combines with the subjectivity of hydrologic interpretation to allow each side of the border to develop divergent models of the aquifer. Differences in the analytic models and analytic frames held by each country allow for each country's water-management values to come to the fore: in the absence of a consensus on hydrologic processes and definitions of safe yield, prioritization of water use and sovereignty concerns drive and legitimize unilateral decisions. At present the water-management activities in both countries are a continuation of the status quo; within each country activities are proceeding in line with that country's management paradigms.

In demonstrating how uncertainty bolsters the influence of within-country political factors, our findings problematize the concepts of "equitable and reasonable", "no harm" and "benefit sharing". Each of those concepts is a routine recommendation in the transboundary literature. Each relies upon a common conception across the border that then forms the basis on which the impacts of water-management activities are evaluated. Yet uncertainty in the physical processes allows for multiple models of the system, and thus supports multiple frames or perspectives. Brugnach *et al.* (2008) suggest that recognition of all three different types of uncertainty – imperfect knowledge of the system, inherent variability and multiple frames – are necessary for problem resolution. However, the hydrological literature focuses on better data and better models without explicit recognition of framing differences, while the management literature acknowledges the uncertainties but usually excludes them from their formal analyses.

There is reason to believe that our findings are representative of transboundary groundwater management beyond the Santa Cruz Aquifer. Due to the large quantity of data required to determine flow processes and the subjectivity of interpretation of the available information, groundwater availability and the impacts of groundwater use are frequently uncertain (Kalf and Woolley 2005). In many countries, small amounts of data are available, as water levels and water-abstraction rates have not systematically been measured, nor has hydrogeologic testing been undertaken. Moreover, as Waterbury (1997, p. 281) explains, "riparian claims typically combine incommensurables – human survival, economic growth, national security" and often "countries do not share common measures of what constitutes legitimate demand". Such incommensurable ideas of needs, demands and legitimacy would readily lead to divergent frames within which "the" groundwater situation is to be understood and represented.

There are many reasons besides uncertainty for the prevalence of the status quo and the resistance to collaboration over transboundary waters. Jarvis (2010), citing several authors on groundwater management, concludes that politics will often overcome rational management principles. Hamman (2005) suggests that interests vested in the status quo will work to prevent changes to it unless ongoing disputes reach crisis proportions. Drawing on the protracted history of negotiations on the CALFED Bay-Delta Program, Innes and Booher (2010) find that, even when a degree of water collaboration is achieved, collaborative planning always remains in tension with self-interest. Indeed they argue that collaboration can at once challenge parts of the status quo while protecting other aspects of it (Innes and Booher 2010, p. 114). We do not argue, therefore, that less analytic uncertainty alone will lead to common values for water and compatible analytic frames. We do argue that greater uncertainty, by reducing the perceived value of collaboration, will prolong status-quo driven management regimes. Rather than seeking cooperative solutions that maximize joint benefits or commit allocations or flows of water, groundwater collaboration might best begin through efforts designed to reduce the multiple uncertainties that exist. Where uncertainty cannot be minimized, low-regret steps that allow for adaptation as water-management objectives are re-defined, and as knowledge of hydrologic processes, costs and benefits increases, may serve to open up a pathway for cooperation when (and if) its need becomes clear.

Acknowledgements

Support for this research was provided by the National Science Foundation (Doctoral Dissertation Research Improvement and Graduate Research Fellowship Programs), the Morris K. Udall Foundation, the Soroptimist Foundation Founder's Region and the University of California Institute for Mexico and the United States. We would like to thank the many persons interviewed as part of this study, Professor Nicholas Sitar, Professor W. Michael Hanemann as well as two anonymous reviewers for their valuable feedback and assistance.

Notes

1. For example, Drieschova *et al.* (2011) categorize ex-post uncertainties in international surface-water treaties as either exogenous (related to [1] variability in water quality and quantity, [2] resource vulnerability, [3] internal politics, [4] external politics, and [5] market fluctuations) or endogenous (related to the treaty itself including [1] how the treaty will be implemented, [2] how valid data collected is, and [3] how the treaty will be financed). These uncertainties relate to treaty effectiveness. The impact of ex-ante uncertainty has been more frequently analyzed for non-water related environmental agreements (for example, Young [1994], Helm [1998], Kolstad [2005]).
2. In the Municipality of Nogales, Sonora, 5% of residents have water service 24 hours per day, 60% of residents receive water service 12 hours per day, 30% of residents receive water service four to five hours per day, and the remaining 5% of residents receive water every other day (OOMAPAS, pers. comm. 4 October 2007).
3. See Milman (2009, chapter 2) for more detail.
4. All interviews for this research were conducted by the lead author in Spanish or English.
5. The conceptual model of an aquifer is the schematic representation used to describe the key features – hydrogeologic, hydrologic and human-ecosystem interactions – of the groundwater system.
6. Water shortages are experienced through intermittent water service; see footnote above.
7. The Organismo Operador Municipal de Agua Potable Acantarillado y Saneamiento (OOMAPAS) is the municipally run water utility of Nogales Sonora, and the primary abstractor of water from the Santa Cruz Aquifer on the Mexican Side of the border.
8. The University of Sonora also measured water levels in 11 wells quarterly between 2000 and 2003, however much of the time all but a few of those wells were dry (Padilla 2005).
9. Seismocontrol (1996) reports a total pumping of 28.4 Mm^3/year for 1997 and the CONAGUA well census reports a total of 9.77 Mm^3/yr. The CONAGUA "Disponibilidad" document reports 19.6 Mm^3/yr for 1996. (See Milman [2009, table 6.4]).
10. Since this research was completed, additional cross-sections have been developed (CONAGUA 2009).
11. Transmissivity is a measure of the rate at which water moves through an aquifer and is equal to conductivity multiplied by depth.
12. Beven (1993) uses the term equifinality to describe how, given the large number of parameters and the complexity of flow within groundwater simulation modelling, the same output, that is, the same model results, can be obtained from different combinations of model parameters and even from different model structures.
13. These include the gila top minnow, the southwest fly catcher and the yellow cuckoo (United States Fish and Wildlife Service, pers. comm., 12 October 2007).
14. See http://www.uawater.arizona.edu/grants/grants07.html.
15. See http://www.uawater.arizona.edu/documents/Fellowship200607/McCoy.pdf.
16. See http://www.usbr.gov/lc/phoenix/programs/progdescon.html.
17. See http://www.nadbank.org/pdfs/state_projects/FS%20Nogales%20SON%20W.pdf. for information on Phase I.
18. An anonymous reviewer for this journal suggested that these diverging priorities might create a suitable opportunity for side payments. Side payments are commonly considered mechanisms for resolving disagreements, especially in game-theoretic solutions. However, in order for side-payments to be effected, both parties must view the proposed trade-off as acceptable. Where ambiguities mean there are no common perceptions of the system under consideration, and where values are incommensurate, side payments may not be practical. Within Arizona there has been discussion of a payment to Mexico that would allow for artificial recharge of the aquifer; however, as this would entail Mexico rescinding sovereignty over a portion of its water

resources, such a solution has not been able to move forward. See Milman (2009, chapter 3) for details.

19. Although Senate Bill 214 authorized US$50 million for the TAAP during the years 2007 to 2010, only US$500,000 was allocated.

References

Anderson, M. and Woessner, W., 1992. *Applied groundwater modeling: simulation of flow and advective transport*. San Diego: Academic Press.

ADWR, 1995a. *Evaluation of proposed water supply project for Nogales, Sonora* Available from: http://virtual.cocef.org/Impacto_Ambiental_Proyectos/ID47/Evaluation_of_Proposed_Water_Supply_Project.pdf [Accessed Fall 2007].

ADWR, 1995b. *Santa Cruz Active Management Area third management plan*. Available from: http://www.azwater.gov/azdwr/WaterManagement/AMAs/ThirdManagementPlan3.htm#Santa [Accessed Fall 2007].

ADWR, 1997. *Santa Cruz Active Management Area: management goal and program implementation concept paper (Draft - 1A)*. Nogales. Unpublished manuscript.

ADWR, 2007. *Concepts for the assured water supply rules for the Santa Cruz Active Management Area*. Unpublished manuscript.

Bernauer, T., 2002. Explaining success and failure in international river management. *Aquatic Sciences*, 64 (1), 1–19.

Beven, K., 2000. On model uncertainty, risk and decision-making. *Hydrological Processes*, 14 (14), 2505–2506.

Blatter, J. and Ingram, H., 2001. *Reflections on water: new approaches to transboundary conflicts and cooperation*. Cambridge: MIT Press.

Bredeheoft, J.D., 1997. Safe yield and the water budget myth. *Ground Water*, 35 (6), 929–929.

Bredeheoft, J.D., 2002. The water budget myth revisited: why hydrogeologists model. *Ground Water*, 40 (4), 340–345.

Bredehoeft, J.D., 2005. The conceptualization model problem – surprise. *Hydrogeology Journal*, 13 (1), 37–46.

Brugnach, M., *et al.*, 2008. Toward a relational concept of uncertainty: about knowing too little, knowing too differently, and accepting not to know. *Ecology and Society*, 13 (2) online.

Chamberlin, T.C., 1897. The method of multiple working hypotheses. *Journal of Geology*, 5, 837–848.

CONAGUA, n.d. *Determinacion de la disponibilidad de agua en el Acuifero Rio Santa Cruz, Sonora*. CONAGUA internal document.

CONAGUA, 2009. *Determinacion de la disponibilidad de agua en el Acuifero 2615 Rio Santa Cruz, Estado de Sonora*. CONAGUA internal document.

Corkhill, F., 2006. *Draft. Proposed physical availability criteria for the Santa Cruz AMA with special requirements for hydrologic studies for the Santa Cruz AMA*. ADWR internal memo.

De Rocquigney, E., Devictor, N., and Tarantola, S., 2008. *Uncertainty settings and natures of uncertainty. Uncertainty in industrial practice: a guide to quantitative uncertainty management*. Chichester: John Wiley & Sons, 199–211.

Dombrowsky, I., 2007. *Conflict, cooperation, and institutions in international water management: an economic analysis*. Cheltenham, UK and Northampton, MA: Edward Elgar.

Drieschova, A., Fischhendler, I., and Giordano, M., 2011. The role of uncertainties in the design of international water treaties: an historical perspective. *Climatic Change*, 105 (3–4), 387–408.

Erwin, G., 2007. *Groundwater flow model of the Santa Cruz active management area microbasins international boundary to Nogales International Wastewater Treatment Plant Santa Cruz County, Arizona*. ADWR report. Available from: http://www.azwater.gov/AzDWR/Watermanagement/AMAs/SantaCruzAMA/default.htm [Accessed Spring 2008].

Farias, U., 1993. *Derecho Mexicano de aguas nacionales: legislacion, comentarios y jurisprudencia Mexico*. Mexico: Editorial Porrua, S.A.

Fischhendler, I., 2004. Legal and institutional adaptation to climate uncertainty: a study of international rivers. *Water Policy*, 6 (4), 281–303.

Frey, F.W., 1993. The political context of conflict and cooperation over international river basins. *Water International*, 18 (1), 54–68.

Frisvold, G. and Caswell, M., 2000. Transboundary water management: game-theoretic lessons for projects on the US-Mexico border. *Agricultural Economics*, 24 (1), 101–111.

Hamman, R., 2005. The power of the status quo. *In*: J.T. Scholz and B. Stiftel, eds. *Adaptive governance and water conflict: new institutions for collaborative planning*. Washington, DC: Resources for the Future, 125–129.

Helm, C., 1998. International cooperation behind the veil of uncertainty. *Environmental and Resource Economics*, 12 (2), 185–201.

Iida, K., 1993. Analytic uncertainty and international cooperation: theory and application to international economic policy coordination. *International Studies Quarterly*, 37 (4), 431–457.

Innes, J.E. and Booher, D.E., 2010. *Planning with complexity: an introduction to collaborative rationality for public policy*. New York: Routledge.

IBWC, 1973. *Minute no. 242: Permanent and definitive solution to the international problem of the salinity of the Colorado River*. Mexico, D.F.: International Boundary and Water Commission. Available from: http://www.ibwc.state.gov/Files/Minutes/Min242.pdf [Accessed Fall 2007].

Jarvis, W.T., 2010, Water wars, war of the well, and guerilla well-fare. *Ground Water*, 48 (3), 346–350.

Kalf, F. and Woolley, D., 2005. Applicability and methodology of determining sustainable yield in groundwater systems. *Hydrogeology Journal*, 13 (1), 295–312.

Kolstad, C.D., 2005. Piercing the veil of uncertainty in transboundary pollution agreements. *Environmental and Resource Economics*, 31 (1), 21–34.

Küçükmehmetoğlu, M. and Guldmann, J.-M., 2004. International water resources allocation and conflicts: the case of the Euphrates and Tigris. *Environment and Planning A*, 36 (5), 783–801.

Lemarquand, D., 1976. Politics of international river basin cooperation and management. *Natural Resources Journal*, 16, 883–902.

Liverman, D.M., Merideth, R., and Holdsworth, A., 1997. Climate variability and social vulnerability in the U.S.–Mexico border region: an integrated assessment of the water resources of the San Pedro river and Santa Cruz river basins. Tucson, AZ: Udall Center, USGS.

Milman, A., 2009. Bordering on water management: ground and wastewaters in the US-Mexico transboundary Santa Cruz River Basin. Ph.D. thesis. University of California, Berkeley.

Milman, A. and Scott, C.A., 2010. Beneath the surface: intranational institutions and management of the United States – Mexico transboundary Santa Cruz aquifer. *Environment and Planning C: Government and Policy*, 28 (3), 528–551.

Morehouse, B.J., Carter, R.H., and Sprouse, T.W., 2000. The implications of sustained drought for transboundary water management in Nogales, Arizona, and Nogales, Sonora. *Natural Resources Journal*, 40 (4), 783–817.

Mumme, S.P., 2005. Advancing binational cooperation in transboundary aquifer management on the U.S.-Mexico border. *Colorado Journal of International Environmental Law and Policy*, 16 (1), 77–111.

Nelson, K., 2007. *Groundwater flow model of the Santa Cruz active management area along the effluent-dominated Santa Cruz River Santa Cruz and Pima Counties, Arizona*. ADWR report. Available from: http://www.azwater.gov/AzDWR/Watermanagement/AMAs/SantaCruzAMA/default.htm [Accessed Spring 2008].

Oberkampf, W.L., *et al.*, 2004. Challenge problems: uncertainty in system response given uncertain parameters. *Reliability Engineering & System Safety*, 85 (1–3), 11–19.

Oreskes, N., Shrader-Frechette, K., and Belitz, K., 1994. Verification, validation and confirmation of numerical models in the earth sciences. *Science*, 263 (5147), 641–646.

Padilla, G.T., 2005. *Aplicacion de MODFLOW para la Simulación Hídrogeologica Estacionaría de Una Porción de la Cuenca Del Rio Santa Cruz, Sonora, Mexico*. Unpublished master's thesis, Universidad de Sonora, Hermosillo.

Phillips, D.J., *et al.*, 2006. Transboundary water cooperation as a tool for conflict prevention and broader benefit sharing. Stockholm: Swedish Ministry for Foreign Affairs. Available from: http://www.ifh.uni-karlsruhe.de/people/fenton/Water-Resources/Transboundary%20Water%20Cooperation.pdf [Accessed Fall 2007].

Sadoff, C. and Grey, D., 2005. Cooperation on international rivers: a continuum for securing and sharing benefits. *Water International*, 30 (4), 420–427.

Santa Cruz County, 2004. *Santa Cruz County comprehensive plan*. Available from: http://www.co.santa-cruz.az.us/com_development/pdf/complete-plan-2005.pdf [Accessed Fall 2007].

SAGARPA, 2004. Distrito de Desarrollo Rural No. 140 Magdalena [online]. Secretaría de Agricultura, Ganadería, Desarrollo Rural, Pesca, y Alimentación. Available from: http://www.sagarpa.gob.mx/dlg/sonora/ddrs/cader1401.htm [Accessed Fall 2007].

SeismoControl, S.A. de. C.V., 1995. *Diagnostico de las Condiciones Geohidrologicas de la Zona los Alisos - Santa Cruz, Nogales, Sonora*. Technical report. Mexico, D.F.

SeismoControl, S.A. de. C.V., 1996. *Estudio de Modelacion Hidrodinamica en el Aquifero de Santa Cruz, Nogales Sonora*. Technical report. Mexico, D.F., Contrato No. GAS-016-96.

Shamir, E., *et al.*, 2005. *Generation and analysis of likely hydrologic scenarios for the Southern Santa Cruz River*. HRC Technical Report No. 4. Available from: http://www.hrc-lab.org/projects/projectpdfs/SANTACRUZ_REPORTS/TR4_01-10-06.pdf [Accessed September 2007].

Tsur, Y. and Zemel, A., 1995. Uncertainty and irreversibility in groundwater resource-management. *Journal of Environmental Economics and Management*, 29 (2), 149–161.

UN, 1997. *Convention on the Law of Non-Navigational Uses of International Watercourses*. [online] G.A. Res. 51/229, UN GAOR, 51st Sess., UN Doc. A/RES/51/229 (1997). Available from: http://internationalwaterlaw.org/documents/intldocs/watercourse_conv.html [Accessed 29 July 2011].

United States Senate, 2006. Senate Bill 214, 109th Congress. United States - Mexico Transboundary Aquifer Assessment Act. Available from: http://www.govtrack.us/congress/bill.xpd?bill=s109-214 [Accessed September 2007].

Waterbury, J., 1997. Between unilateralism and comprehensive accords: modest steps toward cooperation in international river basins. *Water Resources Development*, 13 (3), 279–289.

Wegerich, K. and Olsson, O., 2010. Late developers and the inequity of "equitable utilization" and the harm of "do no harm". *Water International*, 35 (6), 707–717.

Whittington, D., Wu, X., and Sadoff, C., 2005. Water resources management in the Nile basin: the economic value of cooperation. *Water Policy*, 7 (3), 227–252.

Wilson, J. and Guan, H., 2004. Mountain-block hydrology and mountain-front recharge. *In*: F.M. Phillips, J. Hogan, and B. Scanlon, eds. *Groundwater recharge in a desert environment: the Southwestern United States*. Washington, DC: AGU.

Wolf, A., 1997. International water conflict resolution: lessons from comparative analysis. *Water Resources Development*, 13 (3), 333–365.

Wolf, A.T., 2007. Shared waters: conflict and cooperation. *Annual Review of Environment and Resources*, 32, 241–269.

Young, O.R., 1989. The politics of international regime formation: managing natural resources and the environment. *International Organization*, 43 (3), 349–375.

Young, O.R., 1994. *International governance: protecting the environment in a stateless society*. Ithaca: Cornell University Press.

Zeitoun, M. and Warner, J., 2006. Hydro-hegemony – a framework for analysis of transboundary water conflicts. *Water Policy*, 8 (5), 435–460.

Zhou, Y., 2009. A critical review of groundwater budget myth. Safe yield and sustainability. *Journal of Hydrology*, 370 (1–4), 207–213.

Unitization: a lesson in collective action from the oil industry for aquifer governance

W. Todd Jarvis

Unitization is the joint operation of oil or gas reservoirs by all the owners of rights in the separate tracts overlying the reservoirs. "Pooling" is sometimes referred to as unitization. Unitization as employed in the oil industry is designed to be collectively beneficial. Unitization of aquifers is a pro-market approach that could be used as one means to mitigate the inefficiency of a possession or use-based system of groundwater resources, minimize damage to the storage characteristics of aquifers, promote groundwater exploration and development in underutilized areas and prevent disputes instead of resolving conflicts over the utilization of aquifers.

Unitization certainly did work in the oil and gas context. While it was fought by some, it has proven to be the savior of all.

S.E. Clyde (2011)

Introduction

Building upon the doomsday proclamations about "peak oil" in the popular press and environmental texts, the concept of "peak water" is gaining momentum. The concept of peak oil was first introduced by geologist M. King Hubbert who predicted in 1956 and again in 1968 that oil production from the United States would peak in 1970. Although his predictions were dismissed at the time, US oil production in fact peaked in 1970. (Hubbert 1969, Hall and Day 2009). Intensive exploitation of petroleum "reservoirs" has led to premature depletion and, in some cases, irreversible damage to the storage characteristics of oil and gas reservoirs.

Heinberg (2007) examined other arrays of "peaks" by looking at the bigger picture of the production of global liquid hydrocarbons peaking in 2010, as well as the peaking use of freshwater resources, in his book *Peak everything*. Falkenmark (2009) describes "peak water" as a ". . . new and apt name for the present and looming threat of amplified water scarcity of both blue and green water sources occurring around the world". Gleick and Palaniappan (2010) indicate that water has characteristics of both renewable and non-renewable resources and provide a detailed assessment and definition of three concepts of "peak water": peak renewable water, peak non-renewable water and peak ecological water.

Peak non-renewable water is groundwater stored in aquifers. Considering that groundwater is the world's most extracted raw material, with withdrawal rates approaching 800 to

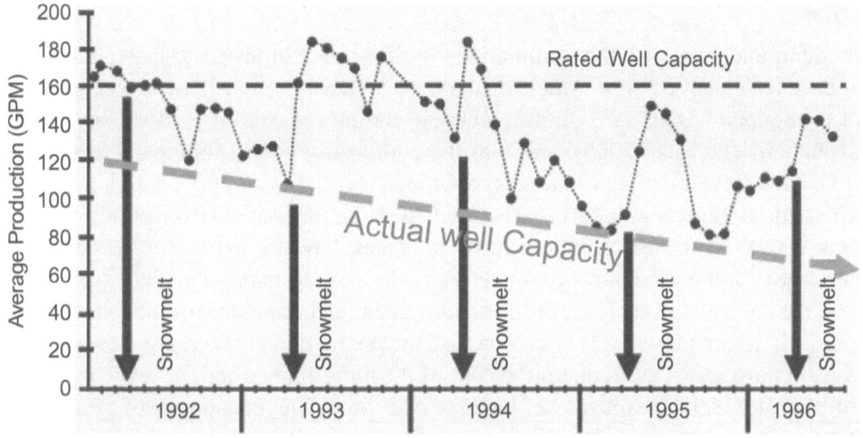

Figure 1. Reduced storage capacity in fractured-rock aquifer. Note: Aquifer performance of fractured-rock aquifer located in Summit County, Utah. Note general decline in well capacity with time, despite apparent recharge during snowmelt, indicative of lost storage due to pumping.

1,000 km^3 per year through millions of water wells, gives some credence to the concerns over peak water (Shah 2009). Reference to "non-renewable groundwater" is controversial, but it usually refers to situations where present-day replenishment is limited but aquifer storage is large, many times "decoupled" from the hydrologic cycle (Foster and Loucks 2006). The concept is not new; the seminal work of C.V. Theis (1940) determined that: "All water discharged by wells is balanced by a loss of water somewhere . . .This loss is always to some extent and in many cases largely from storage in the aquifer . . . Some ground water is always mined."

But peak water is not limited just to mining non-renewable groundwater. Narasimhan (2009) indicates that the definition includes aquifers where the storage characteristics have been permanently changed due to pumping. This phenomenon is limited to confined aquifers: "when the geological material is subjected to unprecedented declines in water pressure (tens of meters of water level decline), it will behave in a 'non-elastic' fashion. Under such non-elastic deformation, a certain amount of porosity (groundwater storage) will be permanently lost (lost storage) when groundwater is extracted at rates causing large declines in water pressure." A national summary of neotectonic subsidence by Holzer (1991) found that neglecting reservoir compressibility in calculating hydrocarbon reserves in rock reservoirs led to estimates that were potentially too large by a factor of two. Jarvis *et al.* (2002) observed a 50% reduction in productivity from a water well targeting a fractured-rock aquifer within just a few years after it was completed; this rock aquifer served as a nearby oil reservoir in Utah (see Figure 1). Narasimhan (2009) reports permanent loss of about 80% of storage space due to non-elastic deformation in three of the large groundwater artesian basins in the United States. Clyde (2011) also recognized that the continued heavy "mining" of water from already over-taxed aquifers would cause subsidence in the formation, thus diminishing the storage capacity of Utah groundwater systems. He correctly points out that once the aquifer storage is damaged, it generally cannot be reversed. Aquifer and oil-reservoir storage permanently lost due to the inelastic properties of the sediments and rocks is non-renewable storage.

Objectives

Lopez-Gunn and Jarvis (2009) summarized the new paradigms for transboundary water management and governance which recognized a tiered approach with levels from the global to the local (Dietz *et al.* 2002), acknowledging the role of civil society and markets (Mukherji and Shah 2005), and that the complexity of environmental systems may require a "post-sovereign" or multi-level approach (Finger *et al.* 2006, Karkkainen 2005). Yet when one considers any "peak" resource situation, economics comes into play (Hall and Day 2009). For example, Vaux (2007) indicates: "Where extractive behavior entails independent action on the part of competitors who seek to maximize their individual net benefits, the aquifer will be managed in a suboptimal and often nonsustainable fashion. In instances where groundwater is not recharged, any extraction is inherently nonsustainable. In these circumstances, the economic theory of the mine applies and the level of extraction should be that which maximizes net benefits over time. The economics of nonrenewable groundwater are identical to the economics of extraction of any nonrenewable resources such as coal, oil and diamonds."

These hydrologic, storage and economic relationships between aquifers and oil reservoirs have direct bearing on groundwater governance and the implementation of the Law of Transboundary Aquifers. Careful consideration of Article 2 of the Law of Transboundary Aquifers detailing the use of terms indicates "*utilization of transboundary aquifers or aquifer system includes extraction of water, heat and minerals, and storage and disposal of any substance*" (emphasis added). Clearly, the new paradigm for groundwater governance extends beyond the equitable and reasonable sharing of water to acknowledging that the available aquifer storage is also a shared resource as described by Puri and Struckmeir (2010).

While the design principles for common-pool resource management work well for the modern geohydrologic balance, unitization provides a more holistic approach to the post-modern geohydrologic balance because the post-modern view of groundwater resources acknowledges not only the pumped groundwater, but also storage of the aquifer system as outlined by Ragone (2007). Consider Figure 2 that compares and contrasts the modern versus post-modern geohydrologic balance. The traditional water balance depicted in the shaded boxes reveals a limited role for aquifer storage: strictly water that is pumped for beneficial use. Yet the post-modern water balance includes sources and sinks that account for changes in socioeconomic forces that add to or displace water that is in storage. The result is more of a market-based approach to utilizing aquifer storage and groundwater.

It is clear the problem of groundwater or aquifer management has morphed from a strictly disciplinary level of understanding to a transdisciplinary paradigm. According to Max-Neef (2005), transdisciplinarity is more than a new discipline but rather a different manner of seeing the world that is more systemic and more holistic. Lawrence (2010) recognized the need to move from a disciplinary to a transdisciplinary framework when a problem is complex and has heterogeneous domains. Transdisciplinarity addresses several levels of reality and through several levels of organization. Max-Neef (2005) describes the different levels ranging from "what exists", incorporating fields of study such as geology and economics, to "what we are capable of doing", that focus on the fields of engineering and commerce. The third level addresses "what we want to do" through planning, law and policy. The top level refers to "what we must do" based on values, ethics and philosophy.

This is the foundation of the concept of "unitization" of oil fields that was developed to protect the "corresponding rights" or "sovereignty" of all pore-space owners in the unit and to not waste valuable pore space. While some readers may consider the concept of unitization another name for common-pool resource management, this paper will provide a

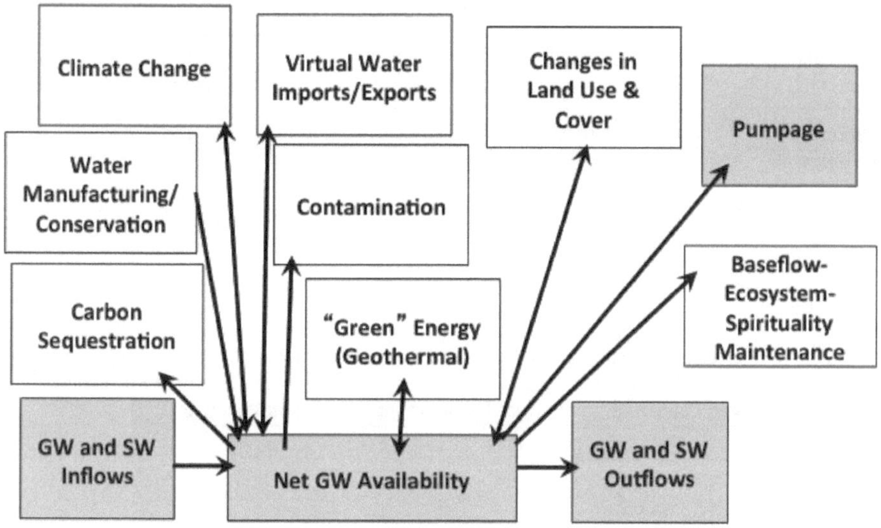

Figure 2. Post-modern geohydrologic balance. Note: Arrows pointing to or away from Net Groundwater Availability indicate gain or loss of water. Double-headed arrows reflect either a gain or loss of water. Shaded boxes indicate components of modern geohydrologic balance. GW = groundwater; SW = surface water. Modified after Ragone (2007).

preliminary comparison between the two concepts. This discussion is followed by a framework arguing for the consideration of unitization as routinely applied in the oil and gas industry in the portfolio of options for groundwater governance, and implementing the Law of Transboundary Aquifers on the bases of what exists, what we are capable of doing, what we want to do and what we must do.

The commons versus unitization

Groundwater and the commons

Groundwater is a resource that is found everywhere. The commons refers to resources, facilities or property institutions with some aspect of joint ownership and access (Buck 1998). Common-pool resources are valued resources that are available to use by more than one person and subject to degradation due to overuse. Groundwater systems are an example of a "pure" common-pool resource because of the costly exclusion and subtractability attributes (Dietz *et al.* 2002).

An exhaustive analysis of common-pool resource theory as applied to groundwater resources is beyond the scope of this paper. The topic has garnered international interest for decades. Groundwater was one of the common-pool resources profiled by the seminal works of Ostrom (1990), Blomquist (1992) and Schlager (2004). On the basis of over two decades of research on the commons, Ostrom (1990) developed eight design principles for the management of common-pool resources as summarized in Table 1. Blomquist (1992) provided a case study of implementing the design principles for the management of groundwater in southern California.

Shah (2009) promotes the importance of "aquifer communities" as a means to "taming the anarchy" of groundwater development by the over 20 million wells in India. India epitomizes a "peak water" situation given that it is the largest groundwater user in the world. Aquifers are being depleted at an alarming rate with approximately 60% of the groundwater "blocks" predicted to be in a critical condition by 2025 (World Bank 2010).

Table 1. Design attributes and principles of unitization versus common-pool resources.

Principle or attribute	Unitization	Common-pool resources
Conceptual development	1890–1930s	1960–90s
Boundaries	• Voluntary units • Compulsory/conservation units • Geographic units • Geologic units	Clearly define boundaries for the user pool and the resource domain
Rules	• Pre-unit agreements at appraisal • Unitization agreement at pre-development • Redetermination during development	Appropriation rules developed for local conditions and provisional rules developed for resource maintenance
Collective action	• Collectively beneficial • Allows sharing of development infrastructure • Avoids unnecessary wells and infrastructure occurring under the competitive rule of capture	Collective-choice arrangements developed by the resource users
Monitoring	• Uses pressure maintenance on the reservoir • Uses best technical or engineering information • Provides foundation to carry out a secondary recovery programme	Monitoring programmes developed for the resource
Sanctions	• Gives all owners of rights in the common reservoir a fair share of the production	Graduated sanctions developed for "violators" of the rules
Dispute resolution	• Pre-unit agreement • Industry-standard agreements • Redetermination process	Conflict-management schemes developed
Rights of regimes	• Can be developed through voluntary or government-mandated compulsory action	Rights of organized environmental regimes respected by external authorities
Administration	• Voluntary to compulsory • Other alternatives (for example, sole development, partitioned development, fixed equity, buy-out, or asset swaps)	Nested enterprises used to administer management

The World Bank (2010) completed The Andhra Pradesh Drought Adaptation Initiative (APDAI), a pilot project that implemented the design principles of common-pool resource management in India. Before the APDAI project, only the richer farmers had access to groundwater because they could afford to dig deep wells. The poor farmers had to depend on the unreliable monsoon rains to irrigate their crops. But according to Narenda (2010), ". . . convincing the richer farmers to share the water from their wells was not easy. However they agreed to do so because many of them too had fields that were far from their wells. If a pipeline was laid, these fields would also gain access to water." The project's success can be attributed to some of the design principles for community management of common-pool resources.

Aquifers and unitization

In contrast to the psychology of the commons that focuses more on the individual operating in an environment of a common-pool resource, it is useful to explore the psychology of unitization as it emphasizes the collective in the management of the commons. Cognitive psychologist Robert Goldstone (2000) defined unitization as a single functional unit that is created for a complex pattern. This functional unit can be identified without an analytic process of breaking it down into components and identifying the components. For example, examination of Figure 3 reveals that if a group of shapes, chickens in this case, form a coherent, contiguous pattern and are often repeated together, a single chunk or unit will be formed by joining or linking them together. Goldstone argues that unitization is the opposite of dimensionalization and segmentation, common attributes of the commons. Unitization creates a single unit from multiple parts that occur together.

The reader is referred to Libecap and Wiggins (1985), Libecap and Smith (2001) and Worthington (2011) for excellent summaries of unitization theory as applied in the US oil industry. Unitization is defined as the joint operation of an oil or gas reservoir by all the owners of rights in the separate tracts overlying the reservoir or reservoirs (Weaver and Asmus 2006, Bastida *et al.* 2007). "Pooling" is sometimes referred to as unitization. While oil exploration and production are usually regarded as private property, the production of oil and gas has always been treated as affecting the public interest because it has beneficial use to both the private and public welfare and is an important revenue stream for the many states and countries (Knowlton 1939).

Without unitized operation of the reservoir, the common law "race to the pumps" results in competitive drilling and production with consequent economic waste and physical

Figure 3. Conceptual model of unitization.
Source: Diagram used by permission of Cognitive Science Program, Indiana University.

damage to the pool and reservoir, with each owner drilling more and pumping faster than their neighbour. The concept was designed to permit reservoir engineers to plan operation of an oil "pool" in order to conserve petroleum resources. To dispel the myth that unitization is part of the theory on common-pool resources, it is important to note the early history of unitization is centred almost entirely around a pioneer in the US oil industry, Henry L. Doherty, who apparently coined the term and promoted the concept in the early 1890s (Knowlton 1939). While the institution of unitization started in the United States, it is spreading to other parts of the world such as Angola, Azerbaijan, Brazil, China, Colombia, Ecuador, Egypt, Indonesia, Nigeria, Russia, United Kingdom and Yemen (Weaver and Asmus 2006).

Units generally are classified by purpose and are acknowledged as the best method of producing oil and gas efficiently and fairly. On the basis of over 100 years of application, what we are capable of doing has been summarized by Knowlton (1939), Weaver and Asmus (2006), Bastida *et al.* (2007) and Worthington (2011), who conclude that the purposes and advantages of unitization include the attributes listed in Table 1. There is a broad spectrum of oil units recognized by legal scholars depending where the units are used, but they are of four general types as summarized by Kumar (2007):

(1) *Voluntary units*: agreements among interested parties that can be undertaken for exploration or conservation;
(2) *Compulsory or conservation units:* a high level of knowledge of physical characteristics of the "pool" is required for conservation;
(3) *Geographic units:* typically applied where poor well-control precludes defining boundaries of productivity, or in areas of complex geology;
(4) *Geologic units:* typically applied based on geology, productive area, lease position, precedent in a field, producing horizon or trend or economics.

In the case of groundwater, what we want to do has been suggested by Libecap (2005) where "government-mandated unitization of groundwater . . . is a solution to excessive access and drawdown . . . [where] a single 'unit operator' extracts from and develops the reservoir. All other parties share in the net returns as share holders." No case studies exist regarding unitization of aquifer systems, but the principles of unitization are being applied in a few settings. In 2010, farmers and ranchers in the Escalante Valley of southern Utah located in the southwestern United States were facing water-level declines over 30 metres that occurred in the past 50 years. Lost aquifer storage was revealed through neotectonic subsidence: earth fissures tens of metres in length and up to two metres in depth. While the State of Utah introduced a plan to reduce groundwater use by 90%, the water users found the plan unacceptable and "pooled" their water rights to share in reductions in water use by voluntarily forming a unit – the Escalante Valley Water Users Association. The State of Utah passed legislation transferring the management of groundwater to the district that covers two counties (Hansen 2010).

Clyde (2011) describes how the unitization process was considered in the critical groundwater basins in the Milford Flat area in western Utah. The proposed process mimicked that used in the oil and gas industry where:

(1) the State Engineer would need statutory authority to enforce groundwater unitization;
(2) a "watermaster" or common unit operator such as a public district, irrigation district or a non-profit corporation formed by the community, would have authority to regulate the pumping of groundwater and use by everyone in the unit; and

(3) with technical guidance from the State Engineer who would operate like a state or international oil and gas conservation commission, decisions on production and steps to be taken to protect the reservoir would be made by all owners within the unit.

Clyde (2011) reported that some legislators thought the unitization approach to groundwater management ". . . was too utopian and questioned whether such a system of groundwater management could work". Knowlton (1939) reported the same pushback from the oil industry in the early days of unitization when he stated: "We have conquered Mother Nature; now we have only to conquer human nature."

The issue of collective action in the management of surface water and groundwater has been under implementation for decades primarily under the precept of participatory irrigation management (PIM) through water users' associations (WUAs) as summarized by Groenfeldt and Svendsen (2000). "Units" as defined by PIM are small irrigation systems managed by users (farmers) who collectively organize as "hydrological units" (for example, left and right main canals) and decide on surface water or groundwater availability to meet the starting and ending dates of the irrigation season, as opposed to leaving these decisions to larger irrigation districts. However, units as defined by PIM should not conflate sharing of groundwater among the users with sharing the aquifer. Unitization as used in this paper does not address sharing of groundwater, per se, but rather the aquifer: the storage capabilities of the aquifer used for groundwater extraction, manufactured water storage, energy extraction and waste storage. A good example of the differences is that the "units" defined by PIM are simply sharing the resource – the water – which could be akin to sharing catches of fish from a shared river, lake or the ocean. The "units" as defined through unitization are determining not only what can be extracted (or captured) but also the storage space of the aquifer, which would be akin to focusing on the habitat where the fish reside. PIM focuses on the modern hydrologic balance of acknowledging that aquifers simply store water for pumping; unitization focuses on the post-modern hydrologic balance recognizing that aquifers have many purposes beyond serving as storage containers for water.

Knowlton (1939) recognized three principal fronts upon which the problem of unitization could be attacked in the early stages of oil unitization in the United States. For groundwater, the least difficult units to form will be the ones addressing the problem of underutilization of groundwater as described by Giordano (2009) or deeper, undeveloped aquifer systems such as the "megawatersheds" described by Bisson and Lehr (2004). In these cases it should not be difficult to develop the means of establishing and dividing equities and benefits.

Conversely, it is difficult to unitize an area undergoing intensive exploitation. At this stage there is only a limited amount of information available to all parties due to information asymmetry; however, Knowlton (1939) long ago recognized that agreements can be reached ". . . by the broadmindedness and generous approach to the problem by those who are conscious that the benefits received from unit operations will far offset any loss which might be suffered by their failure to receive their accurate equitable share". However, Clyde (2011) also suggested that unitization could also be applied in urbanized areas where municipalities compete for common water supplies. He points to the creation of large urban water districts that has accomplished much towards a unitization plan when addressing large new water developments.

The concept can be further extended to the new strategies to combat climate change by storing carbon in aquifers. For example, the State of Wyoming passed HB 80, which became law on 1 July 2009, and provides for carbon-sequestration units. Likewise, the National Energy Technology Laboratory (2010) reports that several US states use a consistent model that would vest ownership of the subsurface pore space to the owners of the overlying surface. North Dakota and Montana recently adopted similar legislation.

Implementation of the US Energy Policy Act of 2005 resulted in the unitization of US Bureau of Land Management geothermal leases. The purpose of the unitization process is to coordinate the efficient, productive development of multiple geothermal leases in order to protect the public interest by preventing waste or inefficient use of the resource and to ensure maximization of revenues. As geothermal resources are considered "water" resources in some US states, state law vests ownership of geothermal rights with water rights. For example, the Utah Geothermal Resource Conservation Act described under 73-22-7 describes the cooperative or unit operation of a geothermal area.

Shimada (2008) provides an interesting case study balancing the spiritual value of water with private property rights using the concepts of unitization in Kumamoto, Japan. The city water is 100% supported by groundwater. Under Japanese law, groundwater resources belong to the landowners. Drilling data in the Kumamoto area permitted mapping with a high degree of certainty in a region lacking a confining layer separating the unconfined and confined aquifers. The absence of a confining layer in this area permitted rainwater and irrigation water to recharge directly into the principal aquifer system. The groundwater recharged in this area flowed to springs filling Lake Ezu in Suizenji Park, a spiritual shrine; the spring flows have diminished 20% in the past decade. The recharge area is overlain by paddy fields that lie outside the jurisdictional boundaries of Kumamoto. Kumamoto pays the local farmers who voluntarily fill their abandoned paddies within the identified "unit" to recharge the aquifer.

Examples of "privatizing" groundwater include the unintentional poisoning of groundwater by agricultural and industrial wastes that are manifold in nearly every country; unitization of contaminated groundwater may provide new opportunities to remediate and market the previously unusable water and storage. Likewise, unitization may promote increased private investment in manufactured water (for example, aquifer storage and recovery) much like the enhanced recovery methods used in the oil industry.

Clearly, there are many similarities to the design principles for the management of common-pool resources and the principles of unitization. Yet important differences are also obvious: (1) community management of common-pool resources focuses more on the resource (groundwater) than the preserving or managing of the storage characteristics of the reservoir; and (2) unitization focuses more on dispute prevention rather than the conflict-management schemes through minimizing well interference issues and associated costs and through compulsory unitization if voluntary efforts to unitize fail.

Social psychologist Mark van Vugt (2009) identified four conditions or "what we must do" for the successful management of shared resources: information, identity, institutions and incentives. Clearly, transboundary aquifer situations can create competition and conflict between communities and institutions that do not promote the welfare of groundwater resources. Van Vugt (2009) suggests that it is important to think of ways to "blur" the boundaries by promoting that "we are all in this together" by acknowledging the value-added benefits of preventing disputes and promoting efficient management of shared resources. Knowlton (1939) promoted this philosophy when recommending that unitization agreements be entered into "with an open mind. Although we may feel, when the deal is consummated, that we have not received the share in the unit which the value of our properties warranted, we also know that because we have been willing to cooperate we are now the owner of an interest in a unit which will pay far more ultimate profit than our individual holdings would have netted". The core principles, or "4P" framework, behind unitization of transboundary aquifers include:

(1) Promote groundwater exploration and development in underutilized areas, for example, in "megawatersheds" that are being promoted as a new exploration paradigm;
(2) Preserve the storage capacity of aquifers by promoting local control of groundwater development;
(3) Promote private investment in the "post-modern hydrologic balance" including aquifer storage and recovery (ASR), managed recharge (similar to secondary and tertiary recovery operations used in the oil and gas industry), non-renewable groundwater which does not fit well within the paradigm of integrated water-resources management, as well as other opportunities such as remediating contaminated groundwater, ecosystem services, and the spirituality of water;
(4) Prevent disputes by "blurring the boundaries", thus creating a new community of users with a superordinate identity who agree to how to "share" groundwater and aquifers.

Unitization can serve as one tool in implementing the Law of Transboundary Aquifers by not only acknowledging equitable and reasonable use of groundwater, but also by recognizing the sovereignty of each aquifer state by valuing the pore space. According to Worthington (2011), cross-border unitization of oil and gas reservoirs is possible where international boundaries are not disputed. International unitization usually dictates two levels of governance: (1) the states involved need to legislate a field-exploitation agreement for the reservoirs; and (2) a unitization agreement needs to be developed.

Conclusions

Groundwater (aquifer) management has evolved from disciplinary understanding to a transdisciplinary paradigm. The Law of Transboundary Aquifers redefined the geohydrologic balance with a post-modern viewpoint that includes extraction of water, heat and minerals, and storage and disposal of any substance. As a consequence, collective action of aquifer governance is transdisciplinary, and must include integrating "what exists" with "what we are capable of doing" and "what we want to do" to "what we must do". Unitization has been applied by the oil and gas industry for over 100 years and embraces the transdisciplinary paradigm. Unitization provides a framework to address many other situations and benefits associated with aquifers beyond the traditional focus of groundwater allocations from aquifers.

Acknowledgements

This paper was originally presented at the December 2010 "Transboundary aquifers: challenges and new directions" (ISARM2010) international conference sponsored by the United Nations Economic, Social and Cultural Organization (UNESCO) in Paris. The paper benefited greatly from discussions with Ebel Smidt of SG Consultancy and Mediation Ltd and PhD Fellow of Delft Technical University. Dr Robert Goldstone, Chancellor's Professor of Psychological and Brain Sciences and Director of the Cognitive Science Program, Indiana University, graciously provided permission to use the conceptual diagram on unitization. John W. Jarvis provided importance guidance on comparing and contrasting unitization concepts between oil and water. The author acknowledges two anonymous reviewers for their careful reading and helpful comments, especially related to units as used in participatory irrigation management.

References

Bastida, A.E., *et al.*, 2007. Cross-border unitization and joint development agreements: an international law perspective. *Houston Journal of International Law*, 29 (2), 355–368.

Bisson, R.A. and Lehr, J.H., 2004. *Modern groundwater exploration: discovering new water resources in consolidated rocks using innovative hydrogeologic concepts, exploration, aquifer testing, and management methods*. Hoboken, NJ: Wiley Interscience.

Blomquist, W.A., 1992. *Dividing the waters: governing groundwater in southern California*. San Francisco: ICS Press.

Buck, S.J., 1998. *The global commons: an introduction*. Washington, DC: Island Press.

Clyde, S.E., 2011. Beneficial use in times of shortage: respecting historic water rights while encouraging efficient use and conservation. *The Water Report*, 83, 1–13.

Dietz, T., *et al.*, 2002. The drama of the commons. *In:* E. Ostrom, *et al.*, eds. *The drama of the commons*. Washington, DC: National Research Council, National Academy Press, 3–35.

Falkenmark, M., 2009. Peak water. Entering an era of sharpening water shortages [online]. *Stockholm Water Front*. 10–11 December, Available from: http://www.siwi.org/documents/Resources/Water_Front/WF-3-4-2008-lowres.pdf [Accessed 21 July 2011].

Finger, M., Tamiotti, L., and Allouche, J., 2006. Introduction: conceptual elements. *In*: M. Finger, L. Tamiotti, and J. Allouche, eds. *The multi-governance of water: four case studies*. Albany: State University of New York Press, 1–42.

Foster, S. and Loucks, D.P., eds, 2006. *Non-renewable groundwater resources. A guidebook on socially sustainable management for water-policy makers*. IHP-VI, Series on Groundwater 10. Paris: UNESCO.

Giordano, M., 2009. Global groundwater? Issues and solutions. *Annual Review of Environment and Resources*, 34, 153–178.

Gleick, P.H. and Palaniappan, M., 2010. Peak water: conceptual and practical limits to freshwater withdrawal and use [online]. *Proceedings of the National Academy of Sciences (PNAS)*, 107 (25), 11155–11162. Available from: http://www.pacinst.org/press_center/press_releases/peak_water_pnas.pdf [Accessed 21 July 2011].

Goldstone, R.L., 2000. Unitization during category learning. *Journal of Experimental Psychology: Human Perception and Performance*, 26 (1), 86–112.

Groenfeldt, D. and Svendsen, M., eds, 2000. *Case studies in participatory irrigation management. WBI Learning Resource Series*. Washington, DC: The International Bank For Reconstruction And Development/The World Bank.

Hall, C.A.S. and Day, J.W., Jr., 2009. Revisiting the limits to growth after peak oil [online]. *American Scientist*, 97 (3), 230. Available from: http://www.pelicanweb.org/solisustv05n06page2halldayamsci.html [Accessed 21 July 2011].

Hansen, J., 2010. It takes a district: Utah landowners control groundwater use Escalante Valley citizens plan to save their declining aquifer [online]. *High Country News*. 10 May. Available from: http://www.hcn.org/issues/42.8/it-takes-a-district [Accessed 21 July 2011].

Heinberg, R., 2007. *Peak everything*. Gabriola Island, BC: New Society Publishers.

Holzer, T.L., 1991. Neotectonic subsidence. *In*: G.A Kiersch, ed. *The heritage of engineering geology: the first hundred years. Centennial special volume 3*. Boulder: Geological Society of America.

Hubbert, M.K., 1969. Energy resources. *In*: The National Academy of Sciences–National Research Council, Committee on Resources and Man, eds, *Resources and Man – a study and recommendations*. San Francisco: W.H. Freeman, 157–242. Available from: http://www.hubbertpeak.com/hubbert/ResourcesAndMan.pdf [Accessed 21 July 2011].

Jarvis, W.T., Yonkee, A., and Matyjasik, M., 2002. Sustainability of fractured bedrock aquifers. Implications for growth management policy, Summit County, Utah. *In*: *Proceedings of 2002 Annual Meeting*, American Institute of Hydrology, 13–15 March 2002, Denver, Colorado. Available from: http://info.ngwa.org/GWOL/pdf/020778053.pdf [Accessed 21July 2011].

Karkkainen, B.C., 2005. Transboundary ecosystems governance: beyond sovereignty? *In*: C. Bruch, *et al.*, eds. *Public participation in the governance of international freshwater resources*. Tokyo: United Nations University Press, 73–87.

Knowlton, D.R., 1939. Unitization-Its progress and future. *Drilling and Production Practice, American Petroleum Institute Report*, paper 39, 630–635.

Kumar, M.B., 2007. Geological aspects of unitization in the petroleum fields of Louisianna: a brief overview. *The Professional Geologist*, 30–32. Available from: http://dnr.louisiana.gov/assets/docs/TPG_Kumar-Articles.pdf [Accessed 21July 2011].

Lawrence, R.J., 2010, Beyond disciplinary confinement to imaginative transdisciplinarity. *In*: V.A. Brown, J.A. Harris, and J.Y. Russell, eds. *Tackling wicked problems through the transdisciplinary imagination*. Washington, DC: Earthscan, 16–30.

Libecap, G.D., 2005. *The problem of water*. Essay prepared for National Bureau of Economic Research. Available from: http://www.aeaweb.org/annual_mtg_papers/2006/0108_1300_0702.pdf [Accessed 4 June 2011].

Libecap, G.D. and Smith, J.L., 2001. Regulatory remedies to the common pool: the limits to oil field unitization. *The Energy Journal*, 22 (1), 1–26.

Libecap, G.D. and Wiggins, S.N., 1985. The influence of private contractual failure on regulation: the case of oil field unitization. *Journal of Political Economy*, 93 (4), 690–714.

Lopez-Gunn, E. and Jarvis, W.T., 2009. Groundwater governance and the law of the hidden sea. *Water Policy*, 11 (6), 742–762.

Max-Neef, M.A., 2005. Foundations of transdisciplinarity. *Ecological Economics*, 53 (1), 5–16.

Mukherji A. and Shah, T., 2005. Socio-ecology of groundwater irrigation in South Asia: an overview of issues and evidence. *Hydrogeology Journal*, 13 (1), 328–345.

National Energy Technology Laboratory, 2010. Best practices for site screening, selection, and initial characterization for storage of CO2 in deep geologic formations. DOE/NETL-401/090808 [online]. Available from: http://www.netl.doe.gov/technologies/carbon_seq/refshelf/BPM-SiteScreening.pdf [Accessed 21 July 2011].

Narasimhan, T.N., 2009. Groundwater: from mystery to management. *Environmental Research Letters*, 4 (July–September 2009) 035002 doi:10.1088/1748-9326/4/3/035002.

Narenda, C., 2010. India: An innovative way of sharing diminishing groundwater [online]. Available from: http://www.mynews.in/News/india_an_innovative_way_of_sharing_diminishing_groundwater_N41355.html [Accessed 4 June 2011].

Ostrom, E., 1990. *Governing the Commons: the evolution of institutions for collective action*. Cambridge: Cambridge University Press.

Puri, S. and Struckmeier, W., 2010. Aquifer resources in a transboundary context: A hidden resource? – Enabling the practitioner to "see it and bank it" for good use. *In*: A. Earle, A. Jägerskog, and J. Öjendal, eds. *Transboundary water management: principles and practice*. Earthscan: London, 73–90.

Ragone, S.E., 2007. The post-modern water balance and its role in groundwater management. *In*: S. Ragone, *et al*., eds. *The global importance of groundwater in the 21st Century*. Westerville, OH: NGWA Press, 119–127.

Schlager, E., 2004. Common-pool resource theory. *In*: R.F. Durant, D.J. Fiorino, R. O'Leary, eds. *Environmental governance reconsidered: challenges, choices, and opportunities*. Cambridge, MA: The MIT Press, 145–175.

Shah, T., 2009. *Taming the anarchy: groundwater governance in South Asia*. Washington, DC: Resources for the Future Press.

Shimada, J., 2008. Sustainable management of groundwater resources for over 700,000 residents in Kumamoto area, Japan. *Proceedings of Symposium on Integrated Groundwater Sciences and Human Well-being*, 36th IAH, Toyama, Japan, 104–111.

Theis, C.V., 1940. The source of water derived from wells-essential factors controlling the response of an aquifer to development, *Civil Engineering*, 10 (5), 277–280.

van Vugt, M., 2009. Triumph of the commons: helping the world to share. *New Scientist*, 2722, 40–43.

Vaux, H., Jr., 2007. The economics of groundwater resources and the American experience. *In*: S. Ragone, *et al*., eds. *The global importance of groundwater in the 21st Century*. Westerville, OH: NGWA Press, 167–176.

Weaver, J.L. and Asmus, D.F., 2006. Unitizing oil and gas fields around the world: a comparative analysis of national laws and private contracts. *Houston Journal of International Law*, 28 (1), 3–103.

World Bank, 2010. *Deep wells and prudence: towards pragmatic action for addressing groundwater overexploitation in India*. Washington, DC: The International Bank for Reconstruction and Development/The World Bank.

Worthington, P.F., 2011. Contemporary challenges in unitization and equity redetermination of petroleum accumulations. *SPE Economics & Management*, 3 (1), 10–17.

Transboundary groundwater law in Europe: a look at an evolving multi-level regime

Götz Reichert

Transboundary groundwater resources in Europe are increasingly governed by a multi-level regime composed of national, supranational and international law. The article provides an overview of essential features of the 2000 Water Framework Directive and the 2006 Groundwater Directive of the European Union. It identifies their pivotal role not only regarding the domestic groundwater law of European Union member states, but also with respect to international law pertaining to transboundary groundwater resources in Europe.

Introduction

In Europe, international cooperation over transboundary freshwater has a long history. Over the decades, riparian states to rivers like the Danube or the Rhine have concluded international agreements. While these conventions initially focused on surface water, they progressively also deal with groundwater resources.[1] Today, approximately 90 transboundary aquifers have been identified by the United Nations Economic Commission for Europe (UNECE 1999). Consequently, the UNECE is currently preparing an explicatory recognition of the existing provisions relevant to transboundary groundwater under its Convention on the Protection and Use of Transboundary Watercourses and International Lakes (UNECE Convention 2011). In addition to these developments under international law, the European Union (EU) has enacted a comprehensive policy on freshwater since the early 1970s, providing for specific provision on groundwater as early as 1979 (European Council Directive 1979).[2] Since the adoption of the Water Framework Directive 2000/60/EC (WFD) in 2000 (Directive 2000/60/EC 2000) and the supplementing Groundwater Directive 2006/118/EC (GWD) in 2006 (Directive 2006/118/EC 2006), EU water legislation constitutes a legal link between international and national water law, thus forming the nucleus of a multi-level regime for the protection and management of transboundary groundwater resources that is about to evolve on a pan-European level.

Nature and effects of EU groundwater legislation

Due to the unique nature of the European Union, its water related legislation has a major impact on the national water law of EU member states. By conferring exceptionally broad competences to the EU in the "EU Treaties" (Treaty on European Union [TEU] 2009, Treaty on the Functioning of the European Union [TFEU] 2009), its Members States have created an autonomous legal order *sui generis* of "supranational" character:[3] EU law has primacy over the national law of EU member states (European Court of Justice 1964) and can impose obligations on the EU member states and their citizens (European Court of Justice 1963). Of specific relevance for groundwater resources are the competences for the protection of the environment and the prudent and rational utilisation of natural resources (TFEU, Article 191, para. 1). These competences are not limited to issues of an exclusively international character. Therefore, EU legislation on groundwater does not only apply to transboundary resources, but also to those groundwater bodies entirely situated within the territory of an EU member state.

The EU has the power to adopt legislative acts in the form of directives. Although "directives" leave the choice of form and methods for their implementation to the national authorities of the member states, they are legally binding upon the member states regarding the result that has to be achieved. Since directives can be very elaborate, they have the potential to fundamentally shape the domestic legislation of member states. This proves especially true for the WFD, which sets up a comprehensive programme for the progressive development of additional provisions on various aspects of water policy. Therefore, the ongoing implementation of the WFD and its supplementing legal instruments is leading to a far-reaching alignment and harmonisation of the national water law of EU member states.[4]

Water Framework Directive: purpose and scope

The Water Framework Directive is based on the awareness that the different components of the hydrologic cycle are interdependent and interact with terrestrial ecosystems. For the first time, qualitative and quantitative aspects of surface waters and groundwater are regulated in an integrated manner on the EU level. "Groundwater" is defined as all water that is below the surface of the ground in the saturation zone and in direct contact with the ground or subsoil (WFD, Article 2, para. 2). A "body of groundwater" entails a distinct volume of groundwater within an "aquifer", representing a subsurface layer of rock of sufficient porosity and permeability to allow either a significant flow of groundwater or the abstraction of significant quantities of groundwater (WFD, Article 2, paras 11 and 12).

Good groundwater status by 2015

The WFD sets legally binding objectives regarding the environmental quality of different components of the ecosystem (WFD, Article 4, para. 1). As far as groundwater is concerned, member states have to protect, enhance and restore all bodies of groundwater and ensure a balance between abstraction and recharge of groundwater, with the aim of achieving "good quantitative and chemical groundwater status" by December 2015 (WFD, Article 2, para. 20).

Groundwater is in "good quantitative status" if the level of groundwater in the groundwater body is such that the available groundwater resource is not exceeded by the long-term annual average rate of abstraction (WFD, Article 2, paras 27 and 28; Annex V, Section 2.1). The level of groundwater must not endanger the objective of "good status" for associated surface waters or significantly damage associated terrestrial ecosystems. Parameters for the determination of "good groundwater chemical status" (Article 2, para. 25 and Annex V, Section 2.3) are the concentrations of pollutants and the conductivity of the groundwater body. Accordingly, the chemical composition of the groundwater body has to be such that the concentrations of pollutants do not exhibit the effects of saline or other intrusions, do not exceed specific quality criteria set in the Groundwater Directive (see below), and would not result in failure to achieve "good status" for associated surface waters or terrestrial ecosystems.

Groundwater directive: prevention and control of groundwater pollution

The supplementing Groundwater Directive 2006/118/EC (GWD)[5] foresees measures to prevent and control groundwater pollution, including criteria for assessing "good chemical status" (GWD, Articles 3 and 4, Annex I) and criteria for the identification of significant and sustained upward trends and for the definition of starting points for trend reversals (GWD, Article 5 and Annex IV).

In order to assess whether the chemical status of a groundwater body is good, two types of criteria apply. The GWD itself sets EU-wide "groundwater quality standards" regarding specific concentrations of nitrates and pesticides which must not be exceeded (GWD, Article 3, para. 1a and Annex I). With regard to other pollutants, member states had to establish "threshold values" by December 2008 (European Commission 2010), taking into account at least such detrimental substances as the heavy metals arsenic, cadmium and mercury (GWD, Article 3, para. 1b and Annex II).

Member states have to prevent or limit inputs of pollutants into groundwater (GWD, Article 6). With regard to any substances identified as "hazardous", discharges into groundwater are prohibited. The input of other pollutants merely needs to be "limited" so as to ensure that they do not cause deterioration or significant and sustained upward trends in their concentrations in groundwater. Such measures have to take account of established best practices specified in the other EU legislation.

River basin districts

The managerial framework of the WFD is based not on political or administrative boundaries, but on the hydrological "river basin" defined as "the area of land from which all surface run-off flows through a sequence of streams, rivers and, possibly, lakes into the sea at a single river mouth, estuary or delta" (WFD, Article 2, para. 13). The corresponding management unit is the "river basin district" (RBD), which refers to "the area of land and sea, made up of one or more neighbouring river basins together with their associated groundwaters and coastal waters" (WFD, Article 2, para. 15). Groundwaters that do not completely follow a particular river basin are assigned to "the nearest or most appropriate" RBD (WFD, Article 3, para. 1).

For each RBD, an analysis of its natural characteristics, a review of the impact of human activity, and an economic analysis of water use must be undertaken (WFD, Article 5). For those bodies of groundwater identified as being at risk of failing to meet "good groundwater status" by 2015 or crossing the boundary between two or more member states, certain information has to be collected, for example, on rates of water abstraction or the chemical composition of discharges into groundwater bodies. These analyses should provide a sufficient basis both for monitoring (WFD, Article 8) and for a "programme of measures" aimed at achieving the environmental objectives within an RBD (WFD, Article 11 and Annex VI). Each programme of measures must include certain "basic measures" such as the control of groundwater abstraction or of point-source discharges. If basic measures are insufficient, "supplementary measures" like artificial recharges of aquifers have to be implemented. In order to make the programme of measures operational within a river basin district by

2012, all relevant information has to be incorporated in a "river basin management plan (RMBP)" (WFD, Article 13 and Annex VII), which had to be finalized by December 2009.

Conclusion: evolving multi-level regime for transboundary groundwater

A remarkable feature of the WFD is its approach towards the management of transboundary freshwater resources: if a river basin covers the territory of more than one member state, an "international river basin district" has to be established (WFD, Article 3, para. 3). Accordingly, 40 international river basin districts have been identified, covering more than 60% of the territory of the EU (European Commission 2007b). The member states must ensure that the requirements of the WFD and GWD for achieving "good groundwater status" and all programmes of measures are coordinated for the whole of the RBD with the aim of producing a single "international river basin management plan" (WFD, Article 3, para. 4 and Article 13, para. 2). And the WFD goes even further: where a river-basin district extends beyond the territory of the EU, the member states concerned shall "endeavour" to establish appropriate coordination with the relevant non-EU countries and to produce a single river-basin management plan, with the aim of achieving the objectives of the WFD throughout the river-basin district (WFD, Article 13, para. 3). Accordingly, member states sharing a body of groundwater have to coordinate their activities in monitoring, the setting of threshold values and the identification of relevant hazardous substances. They have to ensure that the establishment of threshold values for transboundary groundwater bodies is coordinated among the member states concerned. Where a body of groundwater extends beyond the territory of the EU, the member states concerned shall "endeavour" to establish threshold values in coordination with the relevant non-EU countries (GWD, Article 3, paras 3 and 4). By providing the option of using structures stemming from international agreements (WFD, Article 3, paras 5 and 6), the WFD has found an elegant solution for dealing with the required international coordination. Accordingly, member states may entrust an international river commission established under international law with the application of the WFD within an international RBD. Consequently, international river-basin management plans were developed by December 2009 for the rivers Danube, Rhine, Elbe, Ems, Meuse, Oder and Scheldt within the framework of their respective conventions.[6] In the Danube River Basin district, for example, riparian countries – EU member states and non-member countries alike – have jointly developed the Danube River Basin Management Plan (DRBMP). The DRBMP is composed of river basin management plans and programmes of measures pursuant to the WFD; it has been developed at different levels and consists of three parts, ranging from Part A ("Roof Reports") on issues of basin-wide relevance, to the more detailed Parts B and C on the national or sub-basin level (ICPDR 2009). In this way, the various international river commissions serve as common platforms for the ongoing implementation of the WDF and the GWD on a basin-wide scale. Thus EU water legislation establishes as a legal link between international and national water law. This represents a remarkable fusion of supranational EU law and international law (Reichert 2005a), thereby creating a multi-level legal regime for the management of transboundary groundwater resources in Europe (Reichert 2005b).

Notes

1. See, for example, Convention on Cooperation for the Protection and Sustainable Use of the Danube River of 24 June 1994 (Convention on Cooperation for the Protection and Sustainable Use of the Danube River 1995) and Convention on the Protection of the Rhine of 12 April 1999 (Convention on the Protection of the Rhine 1999).

2. The Water Information System for Europe (WISE) provides information on EU water policies, data and projects. See: http://water.europa.eu/ [Accessed 30 July 2011].
3. For a comprehensive overview of the supranational nature of EU law see Craig and De Búrca (2008, pp. 268 *et seq.* and pp. 344 *et seq.*)
4. For the status of WFD implementation see European Commission (2007a) and European Commission (2009).
5. The Groundwater Directive had to be implemented by the EU member states before 16 January 2009. Pursuant to WFD Article 22, para. 2, the Groundwater Directive 80/68/EEC of 17 December 1979 will be repealed effective 22 December 2013.
6. The international river basin management plans for the Danube, Rhine, Elbe, Ems, Meuse, Oder and Scheldt are available from: http://ec.europa.eu/environment/water/participation/map_mc/map.htm [Accessed 18 July 2011].

References

Convention on Cooperation for the Protection and Sustainable Use of the Danube River, 1995. *Official Journal of the European Communities*, L 342, 12 December 1997, 19 *et seq*.

Convention on the Protection of the Rhine, 1999. *Official Journal of the European Communities*, L 289, 16 November 2000, 31 *et seq*.

Craig, P. and De Búrca, G., 2008. *EU law: text, cases and materials*. 4th ed. Oxford: Oxford University Press.

Directive 2000/60/EC, 2000. Directive of the European Parliament and of the Council of 23 October 2000 establishing a framework for Community action in the field of water policy (WFD). *Official Journal of the European Communities*, L 327, 22 December 2000, 1 *et seq*.

Directive 2006/118/EC, 2006. Directive of the European Parliament and of the Council of 12 December 2006 on the protection of groundwater against pollution and deterioration. *Official Journal of the European Union*, L 372, 27 December 2006, 19 *et seq*.

European Commission, 2007a. Communication COM (2007) 128 of 22 March 2007. *Towards sustainable water management in the European Union. First stage in the implementation of the Water Framework Directive 2000/60/EC*. Available from: http://ec.europa.eu/transparency/regdoc/recherche.cfm?CL=en& [Accessed 30 July 2011].

European Commission, 2007b. Commission staff working document SEC (2007) 362 of 22 March 2007, pp. 16 and 58 (Annex 3: List of river basin districts). A map of all river basin districts is available from: http://ec.europa.eu/environment/water/water-framework/facts_figures/pdf/2007_03_22_rbd_a3.pdf [Accessed 18 July 2011].

European Commission, 2009. Communication COM (2009) 156 of 1 April 2009. *Report in accordance with article 18.3 of the Water Framework Directive 2000/60/EC on programmes for monitoring of water status*. Available from: http://ec.europa.eu/transparency/regdoc/recherche.cfm?CL=en& [Accessed 30 July 2011].

European Commission, 2010. Report C(2010) 1096 of 5 March 2010. *Report in accordance with Article 3.7 of the Groundwater Directive 2006/118/EC on the establishment of groundwater threshold values*, accompanying Commission staff working document SEC (2010) 166 and Annex III (Information on the Groundwater Threshold Values of the Member States). Available from: http://ec.europa.eu/transparency/regdoc/recherche.cfm?CL=en& [Accessed 30 July 2011].

European Council Directive, 1979. Directive 80/68/EEC of 17 December 1979 on the protection of groundwater against pollution by certain dangerous substances.

European Court of Justice, 1963. Judgment of 5 February 1963, Case 26/62, Van Gend & Loos v. Netherlands Inland Revenue Administration, European Court Reports, 2 *et seq*. Available from: http://eur-lex.europa.eu/JURISIndex.do?ihmlang=en [Accessed 30 July 2011].

European Court of Justice, 1964. Judgment of 15 July 1964, Case 6/64, Costa v. ENEL, European Court Reports, 587 *et seq*. Available from: http://eur-lex.europa.eu/JURISIndex.do?ihmlang=en [Accessed 30 July 2011].

International Commission for the Protection of the Danube River (ICPDR), 2009. Danube *River Basin district management plan. Part A: basin-wide overview*. Available from: http://www.icpdr.org/participate/sites/icpdr.org.participate/files/DRBM_Plan_2009.pdf [Accessed 18 July 2011].

Reichert, G., 2005a. *The European Community's Water Framework Directive: a regional approach to the protection and management of transboundary freshwater resources? In*: L. Boisson de Chazournes and S.M.A. Salman, eds. *Water resources and international law*. Leiden and Boston: Martinus Nijhoff Publishers, 429–472.

Reichert, G., 2005b. *Der nachhaltige Schutz grenzübergreifender Gewässer in Europa – Die Entstehung eines völker- und europarechlichen Umweltregimes*. Berlin: Duncker & Humblot.

Treaty on European Union, 2009. Consolidated version. *Official Journal of the European Union*, C 83, 30 March 2010, 13 *et seq*.

Treaty on the Functioning of the European Union, 2009. Consolidated version. *Official Journal of the European Union*, C 83, 30 March 2010, 47 *et seq*.

UNECE, 1999. *Inventory of transboundary ground waters*. Available from: http://www.iwacportal.org/File/downloads/inventory_transboundary_groundwaters.pdf [Accessed 30 July 2011].

UNECE, 2011. Convention on the Protection and Use of Transboundary Watercourses and International Lakes, 2011. Legal Board Eighth Meeting 24–25 February 2011 (Doc. LB/2011/INF.2). *Application of the UNECE Water Convention to groundwater: explicatory recognition of the existing UNECE regulatory language*.

The Agreement on the Guarani Aquifer: a new paradigm for transboundary groundwater management?

Pilar Carolina Villar and Wagner Costa Ribeiro

Transboundary water-resources management is usually characterized by prior existence of conflicts. The Agreement on Guarani Aquifer, signed by Argentina, Brazil, Paraguay and Uruguay, is one of the few treaties signed in a precautionary context. Although the Plata Basin has faced many conflicts regarding surface waters, there are no conflicts in the use of shared groundwater. This paper focuses on analyzing the structure of this agreement and its capacity to prevent future conflicts and deepen the cooperation between the states.

Introduction

The Agreement on the Guarani Aquifer, signed by Argentina, Brazil, Paraguay and Uruguay, is the first agreement for transboundary groundwater developed under the influence of the United Nations (UN) Resolution 63/124: the Law of Transboundary Aquifers. The most significant point of this Agreement is the absence of conflicts over the use of the Guarani Aquifer prior to the preparation of this document despite its spatial extent, water volume and strategic importance. The signing of a treaty in the absence of problems, tensions or conflict over a transboundary natural resource is unusual, although recommended by numerous scholars (Giordano and Wolf 2003) and UN Resolution 63/124.

There is a large body of scientific literature suggesting that conflicts generate cooperation among states (Wolf 2007, Zeitoun and Mirumachi 2008). Such arguments fit the few existing examples of cooperation among countries over transboundary aquifers: (1) the *Convention relative à la protection, à l'utilization, à la réalimentation et au suivi de la nappe souterraine Franco-Swiss du Genevois*, signed in 2008 by France and Switzerland (replacing the previous agreement of 1977); (2) two technical cooperation agreements for monitoring and exchanging data related to the Program of Development of a Regional Strategy for Utilization of the Nubian Sandstone Aquifer System;[1] (3) and a technical cooperation agreement for the establishment of a consultative mechanism for the Aquifer System of northwestern Sahara[2] (Burchi and Mechlem 2004).

All of these agreements came into existence because of the need to reverse a situation with a high likelihood of conflict. By the time the first Geneva aquifer agreement was signed in 1977, the groundwater levels had fallen drastically to the point that several wells went dry (de los Cobos 2010). The Nubian Sandstone Aquifer System (NSAS) and Northwestern Sahara Aquifer System (NWSAS) are considered non-renewable aquifers and are located in desert or semi-arid regions where water is a scarce resource (Abu Zeid and Elrawady 2010, Sappa and Rossi 2010). Conditions of overexploitation and degradation of water quality can already be observed in many parts of both aquifer systems (Bakhbakhi 2006).

That was not the case with the Guarani Aquifer where the promulgation of the agreement did not occur as a result of tensions over transboundary groundwater resources. Prior regional cooperation experiences intended to solve surface-water conflicts in the area, as well as international and national pressure motivated by the Guarani Aquifer System Project, can be identified as factors that induced the agreement.

This paper aims to analyze the legal and geopolitical context that prompted the signing of the Agreement on the Guarani Aquifer and evaluate its potential for preventing future conflicts and deepening cooperation among the countries overlying the Guarani Aquifer. The research was made through the qualitative analyses of primary and secondary sources, including the results of the Guarani Aquifer Project, MERCOSUR documents, international rules related to water resources, the Agreement on the Guarani Aquifer and the existing literature on the subject.

This paper is organized into five sections, including this introduction. The second section presents the characteristics of the Guarani Aquifer System (GAS) and the various projects that contributed to promoting knowledge about the aquifer and cooperation among the overlying countries. The third section discusses the integration of the GAS with MERCOSUR. The fourth section then addresses international water law pertaining to transboundary aquifers and the Agreement on the Guarani Aquifer. Finally, the fifth section presents conclusions and recommendations.

The Guarani Aquifer System (GAS)

The Guarani Aquifer System has been the subject of international cooperation projects. It is the only transboundary aquifer in Latin America that has a treaty signed between the countries through which it extends: Argentina, Brazil, Paraguay and Uruguay. As shown in Figure 1, it covers an area of 1,100,000 km^2 in east-central South America between 12° and 35° south latitude and 47° and 65° west longitude, in the Paraná Sedimentary Geological Basin (GW MATE 2009). The GAS is formed by: "a sequence of sandstone beds of Triassic Jurassic Age – formed by the processes of continental deposition on a Permo-Triassic Jurassic regional erosion surface and are overlain by Cretaceous basalt flows which are almost equally-extensive and exceed 1,000 m thickness in some areas" (GW MATE 2009, p. 4).

The average thickness of the aquifer is 250 m and the volume of water is estimated to be 30,000 km^3, which is equivalent to 100 years of cumulative flow in the Paraná River (GW MATE 2009). The quality of water is good, with low rates of mineralization in most places (OAS 2009, GW MATE 2009). The age of the water varies. In the areas of confinement, the residence time may exceed 10,000 years (OAS 2009, GW MATE 2009).

Studies in the region have improved our understanding of the aquifer in relation to its dynamics, recharge areas and future transboundary problems. Figure 1 shows a division

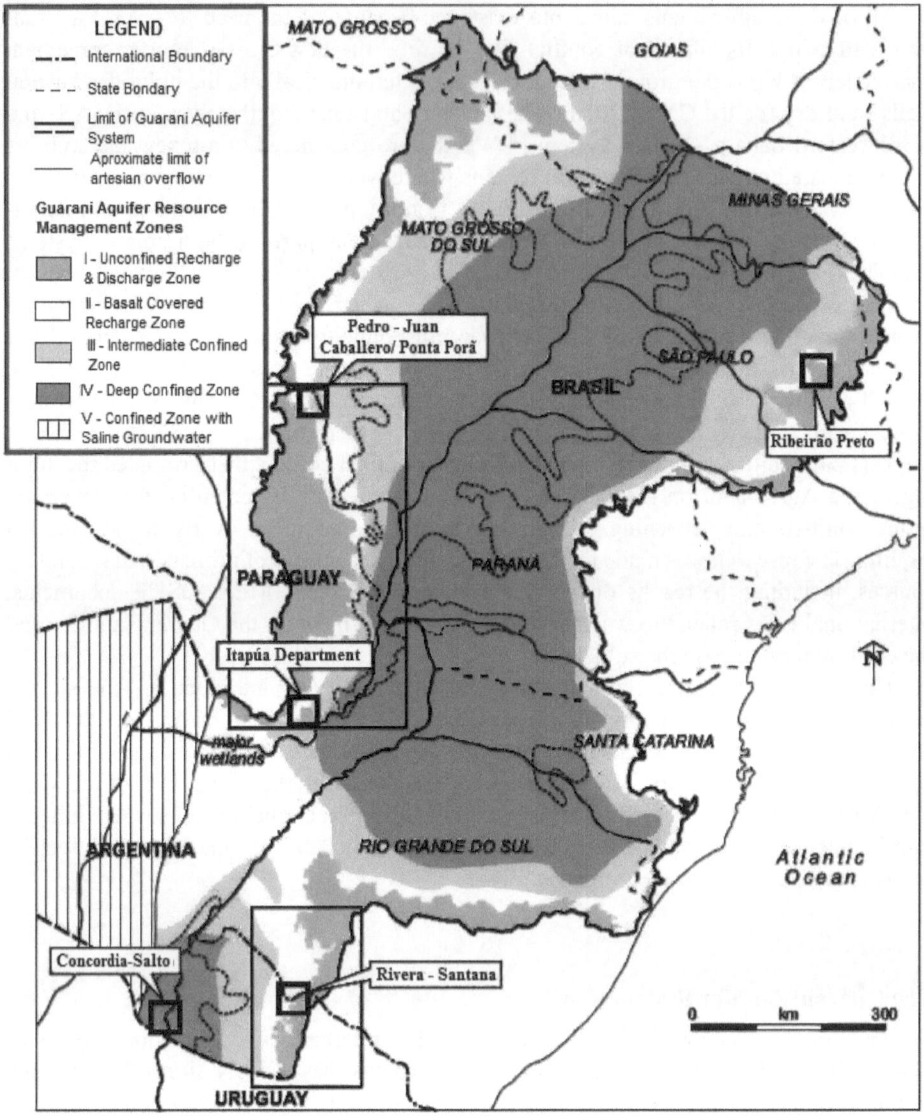

Figure 1. Guarani Aquifer System management zones.
Source: GW MATE (2009, p. 9).

suggested by GW MATE (2009) of the aquifer in areas of management, according to their hydro-geological characteristics:

I – non-confined recharge and discharge zone;
II – recharge zone covered by basalt;
III – intermediate non-confined zone;
IV – deep confined zone; and
V – confined zone with saline groundwater.

This zoning seeks to identify which areas are most vulnerable and enable the pursuit of appropriate management strategies in relation to the particularities of each one. The non-confined recharge and discharge zone (Zone I) is the main source of aquifer recharge and also the most vulnerable to contamination. The recharge zone covered by basalt (Zone II) is overlain by very thick and fractured basalt, which allows recharge, but in a lower capacity than that of Zone I (GW MATE 2009). In the confined zones (intermediate, deep and with a high level of salinity) there is no significant recharge and the extraction of groundwater results in mining the aquifer because there is no water replacement. In contrast, in those areas the aquifer is protected from anthropogenic pollution (GW MATE 2009, OAS 2009).

International cooperation in the Guarani Aquifer System prior to the Agreement on the Guarani Aquifer

International organizations play important roles in the promotion of technical projects between countries. The first joint effort, the Proyecto Sostenible del Acuífero Botucatu[3] (1995), sponsored by the International Development Research Centre (IDRC - Canada), sought to integrate private and governmental institutions in order to establish legal mechanisms and joint measures to manage the aquifer[4] (Borghetti *et al.* 2004).

Another cooperation project was undertaken by the Universidad Nacional del Litoral and Universidad de Buenos Aires, both in Argentina and Universidad de la República in Uruguay[5] (Borghetti *et al.* 2004). This project sought to provide a better understanding of the hydraulic connection between the geological formations that compose the Guarani Aquifer System in both countries.

Paradiplomacy initiatives also contributed to a deeper knowledge of the Guarani Aquifer and the strengthening of its management. An international technical cooperation agreement was signed in 2001 between the Environment Department of the state of São Paulo (Brazil) and the Environment, Public Health and Consumer Protection Department of the state of Bavaria (Germany)[6] (Villar and Ribeiro 2009).

The most important initiative in the region was the Environmental Protection and Sustainable Development of the Guarani Aquifer System Project (also known as the Guarani Aquifer System Project). This six-year (2003–2009) project involved cooperation between the four countries and several international agencies, including the Global Environment Facility (GEF), the World Bank (WB), the Organization of American States (OAS), the Dutch and German Governments and the International Atomic Energy Agency (OAS 2005, 2009).

The project significantly increased knowledge about the GAS's characteristics within the four countries. As a result, conspiracy theories and the climate of distrust[7] that marked the beginning of the project were supplanted by cooperation among stakeholders.

The main objective of the project was to support countries "to elaborate and implement a shared institutional, legal and technical framework to preserve and manage the Guarani Aquifer System (GAS) for the current and future generations" (OAS 2005, p. 1). Four pilot projects were developed to contribute to a better understanding of management aspects and potential conflicts involving these waters.[8]

The conclusions of the studies pointed to an aquifer without transboundary conflicts, a flow of recharge limited to the borders and the basically local nature of any problems. The final document indicated no conflicts over water use among the overlying states, few and localized cases of overexploitation or contamination, and potential sources of conflict restricted to border areas, which represent a small portion of the aquifer (OAS 2009, GW

MATE 2009). The project prioritized the local level stating that this was the appropriate scale at which to manage the Guarani Aquifer:

- "Current and potential transboundary effects of the GAS are restricted to a narrow strip of territory of no more than a few dozen kilometers depending upon local specific hydrodynamic conditions." (OAS 2009, p. 18)
- "Confirmation that groundwater management would be carried out at the local level and that it would involve local stakeholders and water users was, unquestionably, one of the strong points of the Project." (OAS 2009, p. 34).
- "Most pressing and potential groundwater resource management and protection needs of the GAS fundamentally do not have an 'international transboundary' character – albeit that there exist some local 'transboundary hotspots' both between nations, and indeed between individual states of Brazil, that share the aquifer." (GW MATE 2009, p. 14).

Guarani Aquifer management juxtaposes local, provincial, national and international scales, as well as different interests concerning water and soil, which require cooperative management between the different scales and actors. The local level is an important player in groundwater management, but factors at the national and international scales constitute fundamental elements conditioning local decision making; they also have a major role in determining the success or failure of local water politics. The political process of local water policy and management is influenced by many factors and is the result of multiple determinants, many of them external to the process (Harvey 2003).

Studies on the Guarani Aquifer Project promoted the exchange of information among the countries and dialogue among experts, which created a favourable environment for cooperation. Moreover, the convergence of people and national and international funds in the project encouraged scientific production and social attention, which, in turn, resulted in the emergence of an agreement. There were social, academic and international pressures for countries to give some kind of continuity to the project, since it did not attain its initial goal of creating a proposal for a common legal and institutional framework.

In addition, joint efforts promoted by the Guarani Aquifer Project reverberated within MERCOSUR, whose instruments dedicated to environmental issues were used to broaden the debate on the management of the Guarani Aquifer.

MERCOSUR, the environment and the Guarani Aquifer System

International cooperation on the environment between countries of South America is not very common (Ballesteros 2009). The Southern Common Market – MERCOSUR[9] – represents progress for the integration process among the region's countries and has increased cooperation. MERCOSUR was established through the Treaty of Asuncion of 26 March 1991, signed by Argentina, Brazil, Paraguay and Uruguay.[10] The MERCOSUR structure is formed by the Common Market Council[11] (CMC), Common Market Group[12] (GMC), MERCOSUR Trade Commission[13] (CCM), MERCOSUR Parliament[14], Social-Economic Consultative Forum[15] (FCES), and Administrative Secretariat of MERCOSUR.[16] The dispute settlement process was covered by the Protocol of Olivos (2002), which gave new shape to "ad hoc" Arbitral Tribunals and established the Permanent Court of Review. The GMC is assisted in its activities by Working Subgroups, Specialized Meetings and Ad Hoc Groups, among others.

Environmental protection is supported by MERCOSUR and was included in the preamble of its constituent treaty[17] (Ribeiro 2008a). The specialization of environmental issues occurred through the creation of the Specialized Meeting on the Environment, called REMA – Reunión Especializada de Medio Ambiente (GMC Resolution 22/92) – which later gave rise to the Working Subgroup No-6 "Environment" (GMC 20/95).

Subgroup No. 6 pointed out in its Article 6, Item 1, the sustainable management of natural resources, among them water resources. This subgroup has a limited role, because environmental issues are normally subordinated to concerns related to trade and its agenda is set by the GMC (Hochstetler 2003, Ribeiro 2008a). Despite the restrictions, the subgroup made an important contribution to the negotiation of the MERCOSUR Environmental Framework Agreement[18] and the Additional Protocol to the MERCOSUR Environmental Framework Agreement regarding Cooperation and Assistance in Environmental Emergencies[19] (MERCOSUR CMC Decision 14/04) (Hochstetler 2005). Subgroup No. 6 also designed a model Additional Protocol to the Environmental Framework Agreement concerning technical-scientific assistance to the environmental management of transboundary water-resources management.[20]

Despite development of an environmental structure within MERCOSUR, no concrete action has resulted. At the national level, improvements in the policies of member countries have been achieved, but they have experienced considerable difficulties in implementation (Hochstetler 2003).

In addition, Subgroup No. 6 did not participate in the discussions related to Guarani Aquifer, which were conducted by an Ad-Hoc High Level Group specifically created for this subject in 2004 (MERCOSUL / SGT 6 / ATA 4/04). The goal of this group was to formulate a draft shared-aquifer management agreement between the Parties (GMC Decision 25/04 and 48/04). The model suggested was not pursued because of the difficulty in reaching a consensus on a dispute settlement system (Síndico 2010).

This difficulty can be explained by the political context experienced by the block. At the time, the installation of two paper mills on the Uruguay River sparked a conflict between Argentina and Uruguay, which revealed the difficulties and limitations of the countries of the region, the Uruguay River Management Commission (CARU) and MERCOSUR in resolving environmental disputes.

While the paper mills conflict developed, the MERCOSUR Parliament proposed to the CMC the formation of a commission to study, analyze and compare each country's water resource legislation. The goal of the project was to recommend to the national governments modifications to their internal systems (October 2007) and to develop a landmark agreement on the cooperation for the sustainable management of the Guarani Aquifer System (2008). Moreover, the CMC suggested the creation of a regional Research and Development Institute for groundwater and aquifer protection (INRA MERCOSUR - Mercosul/PM/SO/REC. 25/2009), as well as the implementation of a transitional project following the conclusion of the Guarani Aquifer Project (Villar 2010). Unfortunately none of these initiatives has prospered.

In August 2010, during the XXXIX Meeting of the Common Market Council and the Summit of Presidents of the member states of MERCOSUR and Associated States, Argentina, Brazil, Paraguay and Uruguay signed the Agreement on the Guarani Aquifer. Although the MERCOSUR event served as a meeting location to sign the Guarani Agreement, the agreement was developed outside the structure of the MERCOSUR organization. Two reasons can be pointed to as the probable causes of this attitude: the limitations of MERCOSUR in addressing important environmental conflicts in practice, especially the pulp mills case; and the intention of MERCOSUR to include new members that are not located within the Guarani Aquifer System area. The four Guarani countries, which always asserted their sovereign rights over the aquifer, may have considered that an agreement signed under MERCOSUR would allow other countries to interfere with the aquifer's management.

The Agreement on the Guarani Aquifer and the international law of transboundary aquifers

For too long groundwater was not considered a priority in international law. According to McCaffrey: "The law of international groundwater may only be said to be, at best, in the embryonic stages of development" (2001, p. 433). Groundwater resources were addressed indirectly or incidentally as part of surface water, or were simply not mentioned (Mechlem 2003, 2009).

The first legal instrument to mention them was the 1966 Helsinki Rules[21] promoted by the International Law Association (ILA). For many years, this model served as the main instrument representing customary law for shared fresh waters (Salman 2007). The concept of an international drainage basin adopted in the Helsinki Rules included groundwater. However, the peculiarities of these waters demanded that the ILA complement the theme through the 1986 Seoul Rules. Another contribution of this organization are the 2004 Berlin Rules,[22] which revised the legal model of fresh water to include all waters (national and transboundary) and incorporated the products of various fields of international law of recent decades: humanitarian law, human rights and environmental law (Dellapenna 2006). This document established a specific chapter on groundwater. Although the ILA rules have no formal status under international law, it may be argued that they are important and influential instruments in the development of customary international water law (Eckstein and Eckstein 2003)

In 1997, the United Nations General Assembly adopted the United Nations Convention on the Law of the Non-Navigational Uses of International Watercourses. This convention was the result of a long process (McCaffrey 2001), representing the first and only legitimate instrument for countries for the management of shared freshwater (Eckstein and Eckstein 2005). Unfortunately it has not yet entered into force.[23]

Unlike the Helsinki Rules, the Convention departs from the idea of drainage basin and adopts the concept of watercourse, defined in Article 2 as:

> (a) "Watercourse" means a system of surface waters and groundwaters constituting by virtue of their physical relationship a unitary whole and normally flowing into a common terminus;
>
> (b) "International watercourse" means a watercourse, parts of which are situated in different States;

The definition of watercourse expressly includes groundwater, which transforms these waters into a subject of international law. However there are limitations, as Eckstein (2005, p. 5) states: the Convention is restricted to those aquifers that: (1) are physically part of a system of surface and groundwater; (2) are part of a unitary whole; (3) normally flow to a terminus that is common with the hydraulically linked surface water; and (4) have parts of the system located in different states.

The requirement to include groundwater within the scope of the Watercourses Convention, and the related limitations, cause a number of problems. Aquifers can be hydraulically connected to more than one river basin and normally do not flow to a common terminus such as rivers. Strata peculiarities allow groundwater to flow in different directions[24] (sea or other river basins). Also, in many cases, aquifers may not be connected with surface water. Accordingly, the Convention excludes important types of aquifers as those without recharge and those that are supplied solely by rainfall (Mechlem 2003, Eckstein 2005). Due to the complexity of the subject, the ILC intentionally left non-recharging aquifers out of the Convention and preferred to adopt the Resolution on Confined Transboundary Groundwaters,[25] which refers to groundwater resources that are not related to surface waters (Mechlem 2003).

The Guarani Aquifer itself would be excluded by the definition of watercourse considering that except for the areas of recharge, the aquifer is confined by a basaltic layer that impedes any hydraulic connection with surface waters. If the challenge of the last century was to seek to regulate the use of shared surface water, the one of the current century is to consolidate these regulations in order to avoid a water crisis and add a new dimension to the hydrological cycle: groundwater and transboundary aquifers.

The limitations of international law on transboundary aquifers have prompted the International Law Commission of the United Nations to prepare a project on the topic, which was approved on 11 December 2008, by the UN General Assembly, with the title of Resolution 63/124: the Law of Transboundary Aquifers.[26] The approval of Resolution 63/124 symbolizes that the international community recognizes the importance of transboundary aquifers and the need to regulate their use and protection. Although it has no binding legal effect, it represents the first instrument legitimized by the UN General Assembly that is devoted to transboundary aquifers (McCaffrey 2001).

The draft articles contained in the Resolution cover all types of aquifers and considerably broadens the scope of international standards for fresh waters within the United Nations. According to Article 1 of the document, the Draft Articles apply to:

- utilization of transboundary aquifers or aquifer systems;
- other activities that have or may have an impact on aquifers or aquifer systems;
- measures for the protection, preservation and management of aquifers or aquifer systems

Paragraphs (a) and (c), though limited, were already grounded in the UN Water Convention. The novelty lies in subparagraph (b), which may imply a control of land use, somehow aligning itself with the positioning of the 2004 Berlin Rules (Laborde 2010).

The main purpose of this resolution is to encourage states to include transboundary aquifers in their agendas and to establish bilateral and regional agreements on the subject. This resolution has contributed, in terms of political and technical incentives, to the signing of the Agreement on the Guarani Aquifer (Síndico 2010). That occurred shortly after the adoption of Resolution 63/124, on 2 August 2010, under the auspices of the Resolution, the Declarations of the principal global environmental instruments (Stockholm, Rio de Janeiro and Johannesburg), the River Plate Basin Treaty and the Framework Agreement on the Environment in MERCOSUR, when Argentina, Brazil, Paraguay and Uruguay signed the Agreement on the Guarani Aquifer.[27]

It is speculated that this signing was stimulated by three main events: (1) the end of the Guarani Aquifer project, which had provided an important technical foundation; (2) the presentation of Resolution 63/124 on the Law of Transboundary Aquifers by the UN General Assembly; and (3) the International Court of Justice decision of April 2010 in the case of the pulp mills in the Uruguay River between Uruguay and Argentina (Síndico 2010).

The first article of the Guarani Agreement states:

> The Guarani Aquifer System is a transboundary water resource that integrates the sovereign territorial area of Argentina, the Federative Republic of Brazil, the Republic of Paraguay and the Eastern Republic of Uruguay, which are the only holders of this resource and will henceforth be called "Parties".

This treaty sets out the main guidelines of Resolution 63/124 of the United Nations especially in relation to the principles expressed: sovereignty, equitable and reasonable use of water resources, the obligation not to cause harm, cooperation and exchange of data and information.

The agreement reaffirmed the sovereignty of the overlaying states over the aquifer on several occasions. Sovereign territorial domain of the Guarani Aquifer is mentioned in Arts. 1 and 2, while sovereign right over natural resources is set out in Article 3, which states:

> The parties exercise, in their respective territories, the sovereign right to promote the management, monitoring, and sustainable use of water resources of the Guarani Aquifer System and will utilize these resources based on the criteria of rational and sustainable use, and respecting the obligation not to cause appreciable harm to the other Parties or the environment.

This article clearly provides that each country is responsible for managing their portion of the aquifer based on its own public policy, and sets standards for the use and protection of the aquifer. Some scholars argue that this principle merely restates the well-established principle of international law and ensures safeguards for the aquifers, thus preventing the aquifer from being considered as a "common good of mankind similar to the status of a common heritage of mankind" like the deep seabed, outer space or Antarctica (Laborde 2010, p. 3). Other authors view it with surprise and suspicion, contending that the reaffirmation of the sovereignty principle is inconsistent with the spirit of cooperation and equitable use (McCaffrey 2009, McIntyre 2010).

The Agreement follows the main principles and obligations of international law: the equitable use of water resources, the obligation not to cause harm and international cooperation, which includes the exchange of technical information about the aquifer, notification of activities that can cause harm to the parties and obligation to mitigate negative impacts. In this sense, Article 5 states that the Parties to the Guarani Aquifer System should act in conformity with applicable international-law principles and rules if there is a possibility that actions, studies or works within their territory could have effects beyond the national borders.

The principle of equitable use of water resources is affirmed in Articles 3 and 4 and the obligation not to cause harm appears in Articles 3, 6 and 7. In relation to Article 3, it is noteworthy that these principles appear to limit the Parties' sovereign rights related to the Guarani Aquifer System. Each state is responsible for promoting the management, monitoring and sustainable use of the aquifer, but each also has to consider the rational and sustainable use of the aquifer and the obligation not to cause harm.

Cooperation is one of the strong points of the Agreement and appears in many Articles, such as 8, 9, 10, 12, 13 and 14. These statements of cooperation foresee the need for information exchange about water-resource utilization, the duty of prior notification and the right to seek additional information. They also foresee the establishment and development of joint projects and cooperation programmes for technical, scientific and management aspects.

One of the Agreement's weaknesses is that there are no provisions in the document of a specific plan for the protection of recharge areas or for the extraction of non-renewable Guarani waters. Article 14 makes a vague allusion to "identify critical areas, particularly in border areas that require specific treatment measures". The recharge areas could fall into that category, however, the focus of Article 14 seems to be to identify areas that require measures to restrict or control rather than the design of a policy based on precaution and prevention.

Article 15 provides that the cooperation process will be under the responsibility of a Commission formed by the four parties and institutionalized in the mould of Article IV of the River Plate Basin Treaty, which states:

> Notwithstanding the provisions within each country, there will be cooperation bodies and advisory committees of the Government, namely Commissions or National Secretariats, formed in accordance with the Joint Declaration of Buenos Aires. Commissions or Secretariats may establish bilateral contacts, always obeying the criteria and standards of the countries concerned and keeping it properly informed, when appropriate, the Intergovernmental Coordinating Committee.

Thus, Article 15 of the Agreement provides that:

> A Commission composed of the four states party to the agreement will coordinate and cooperate amongst themselves to fulfill the principles and objectives of this Agreement. The committee will promulgate its own regulations.

When implemented, this Commission would be the first to address the issue of transboundary aquifers in the region and in Latin America. At present, it is still not possible to determine the future and scope of this commission. The countries have yet to institute it and endow it with its statutes, powers, membership and budget. Although they already have several international bodies for the management of surface-water resources,[28] whose cooperation promotion scope varies, this Commission would be the first to deal specifically with groundwater, featuring a new form of cooperation in South America. The Commission is a key component of the cooperation process whose success depends on how it will be structured and how it will be supported by the member states.

The resolution of disputes, under Articles 16, 17, 18 and 19 will take place through direct negotiations of the countries, which are obligated to inform the Commission mentioned in Article 15. The role of this Commission is to recommend and give advice to the countries. In accordance with Article 17, if the states fail to reach an understanding, they can request the Commission's recommendations or refer the dispute to arbitration, which will be defined later in an Additional Protocol. The Agreement opted for a new arbitration procedure that is yet to be defined, and ignored the one already provided by MERCOSUR through the Protocol of Olivos.

The initiative of the four countries to establish a joint treaty represents a step forward, given the lack of specific initiatives for transboundary aquifers. However, most of the Agreement statements need domestic regulations and implementation actions. Important points were left out, like the procedural mechanisms for arbitration and strategies for the recharge areas. Despite these imperfections, the signing of the treaty and creation of a Guarani Aquifer Commission represents the intention of the four Guarani countries to deepen the integration process. Cooperation between Latin American countries over transboundary aquifers is a recent phenomenon.

Conclusions

Studies of the Guarani Aquifer System indicate that even though there are limitations in water flows among the countries, the system represents a transboundary water body. The existence of common recharge zones and non-renewable water in the central portion of the aquifer requires the pursuit of shared management solutions in order to rationally exploit and conserve both water and recharge zones. These types of arrangements go well beyond local powers and competence.

The existence of a customs union, such as MERCOSUR, was sidelined in the design of the Agreement. Although we may claim that tools for environmental transboundary cooperation are only emerging, they have actually been present in MERCOSUR almost from its creation. Accordingly, the MERCOSUR structure could be better used in the management of and dispute resolutions concerning transboundary natural resources and environmental cooperation.

The Agreement came into existence during a very active political context. There was the conflict between Argentina and Uruguay over the pulp-mills installation, which was eventually resolved by the International Court of Justice. On the other hand, there was the emergence of UN Resolution 63/124 and the end of the Guarani Aquifer project. All these elements can be interpreted as inducing the development of the Agreement.

The Agreement represents the intention of parties to construct a new paradigm in the sense that it takes place in the absence of groundwater conflict. It is the first time that transboundary groundwater has been considered in a preventive/precautionary context. However, to be really implemented, the preventive and precautionary provisions of the Agreement require domestic and international regulations. The lack of precise cooperation arrangements or an information-exchange process challenges the Agreement's efficiency. Another problem is that the recharge areas, which are the most likely areas to face conflict, are not regulated and there are no strategies for the mining of the aquifer. Finally, the Agreement lacks a solid dispute settlement system to address possible future conflicts.

Because of these issues, the Guarani Aquifer Commission will play a central role in guiding the cooperation process and implementing the Agreement. However, it is still unclear how this commission will be structured by the four countries.

Considering all these limitations, some questions arise. Will the Agreement be implemented or will it be ignored, as MERCOSUR's environmental instruments have been? Will the structures planned be able to strengthen cooperation and resolve future conflicts, or, as happened in the conflict between Argentina and Uruguay, will resolution of disputes rely on organs outside the region? Only time will answer these questions.

Hence the Agreement on the Guarani Aquifer has potential to propose new ways of cooperating and resolving disputes. However, the states of the GAS have only recent undertaken these efforts and, as such, the Agreement requires deeper cooperation initiatives to became effective. The countries will have to structure the Guarani Aquifer Commission, find ways to clearly address central themes that can generate conflict (recharge areas for example) as well as to create means to solve them by the institution of a dispute-settlement system.

Notes

1. Programme for the Development of a Regional Strategy for the Utilization of the Nubian Sandstone Aquifer System was established between Chad, Egypt, Libya, and Sudan. Agreement No. 1 (Terms of reference for the monitoring and exchange of groundwater information of the Nubian Sandstone Aquifer System) and Agreement No. 2 (Terms of reference for monitoring and data sharing) were signed in Tripoli, on 5 October 2000.

2. Establishment of a Consultation Mechanism for the Northwestern Sahara Aquifer System (SASS) signed between Algeria, Libya, and Tunisia at the end of 2002 and the beginning of 2003.
3. Acuífero Botucatu is one of the Guarani Aquifer System geological formations. The denomination Guarani Aquifer was approved by Argentina, Brazil, Paraguay and Uruguay only in May 1996, and intended to include all the different geological formations that are part of the system, as well as honour the Guarani indigenous peoples who live in the area (Borghetti *et al.* 2004).
4. Some of the contributions of this project include: the Technical scientific journey on the management of the Botucatu International Aquifer (1995), the International Workshop on the MERCOSUR Giant Aquifer and the first scientific-technical base of this Aquifer (Borghetti *et al.* 2004).
5. Subsequently, the results of technical cooperation and exchanges between researchers from Argentina and Uruguay were published in Montano *et al.* (1998).
6. This project culminated in a joint project called Pilot System of Information for Environmental Management of Groundwater Resources in the Outcrop Area of the Guarani Aquifer System in the State of Sao Paulo.
7. The fact that the Guarani Aquifer Project was financed by international and foreign institutions fostered a series of unscientific rumours that warned about the vested interests of those institutions and reinforced a discourse of nationalism. These concerns were prompted by the fact that international companies were contracted to prepare major studies of the project, away from the academic centres in the four countries, and the dissemination of some documents only in English, a language not officially spoken in any of the countries.
8. The pilot projects were: (1) Concordia (Argentina) / Salto (Uruguay); (2) Rivera (Uruguay) / Santana do Livramento (Brazil); (3) Encarnación – Cuidad del Este – Caaguazú (Paraguay); and (4) Ribeirão Preto (Brazil). The first two sought a better understanding of issues linked to the management of transboundary groundwater and the others looked for developing management strategies in a groundwater-vulnerability context (presence of recharge areas, risk of contamination or overexploitation).
9. In the second half of the 1980s, Brazil and Argentina began to work more directly to establish a multilateral body that would facilitate and promote trade between countries of South America. This regional experience of integration was embodied in a customs union, called MERCOSUR, for the free movement of goods. MERCOSUR has a legal personality in international law and intergovernmental decision-making capacity. This means that decisions taken in this sphere must be incorporated into domestic legal systems of States Parties to enter into force.
10. In 2006, the Bolivarian Republic of Venezuela signed the Protocol of Accession to MERCOSUR. Bolivia (1996), Chile (1996), Peru (2003), Colombia (2004) and Ecuador (2004) are associate members of MERCOSUR.
11. The Common Market Council is the organ of MERCOSUR that holds the legal personality of the block. It is responsible for the conduct of policy decisions that ensure the achievement of MERCOSUR's objectives. It is composed of the Ministers of Foreign Affairs and Economics, or their equivalents, from each country.
12. The Common Market Group is the executive organ of MERCOSUR. It is composed of 16 members, four from each country. Its assignments are listed in Article 14, XIV of the Protocol of Ouro Preto (POP).
13. The MERCOSUR Trade Commission is designed to assist the Common Market Group and ensure the implementation of instruments that lead to the functioning of the Customs Union (Articles 16, 18 and 21 of the POP).
14. The Parliament of MERCOSUR replaces the Joint Parliamentary Commission – Protocol Establishing the Parliament of MERCOSUR (MERCOSUR/CMC/DEC. 23/05).
15. The Consulting Economic Social Forum is an auxiliary organ representing economic and social sectors. It is composed of an equal number of representatives of the States Parties. Its function is purely advisory.
16. The Administrative Secretariat, based in Montevideo, Uruguay, is headed by a Director, a member of one of the State Parties. The Director is elected for a rotating two-year term by the Common Market Group, appointed by the Common Market Council, and is not eligible for re-election.
17. Treaty for the Establishment of a Common Market between the Argentine Republic, the Federative Republic of Brazil, the Republic of Paraguay and the Republic of

Uruguay (*Tratado de Asunción*). Available from: http://www.mre.gov.py/dependencias/tratados/MERCOSUR/registro%20MERCOSUR/MERCOSURprincipal.htm.

18. This Agreement was approved on 21 June 2001, after 10 years of negotiation and successive revisions. This is the most important environmental legal instrument of MERCOSUR, which reaffirmed the principles already contained in the Declaration of Rio de Janeiro (1992), excluding the precautionary principle. It came into force on 23 June 2004 and is guided by general statements and principles of international environmental law, such as sustainable development, cooperation, public participation and the need to improve and harmonize national environmental regulations.
19. The agreement's main objectives are: to promote cooperation and assistance in cases of emergencies that may jeopardize the environment and population, to harmonize procedures for acting in these cases and to promote the exchange of information. Nevertheless, the instrument has not entered into force.
20. The draft text of this Additional Protocol can be found at: http://www.ambiente.gov.ar/archivos/web/MERCOSUR/File/xxxii anexo11.pdf
21. Helsinki Rules, 1966. Available from: http://www.colsan.edu.mx/investigacion/aguaysociedad/proyectofrontera/Helsinki%20Rules%201966.pdf.
22. Berlin Rules, 2004. Available from: http://www.cawater-info.net/library/eng/l/berlin_rules.pdf.
23. According to Article 36, the Convention needs 35 signatories to enter into force. For up-to-date information see: http://untreaty.un.org/.
24. The case of groundwater in the Danube Basin is a good example. While most of the groundwater in the basin flows parallel to the river toward the Black Sea, a small portion flows in a different direction and serves as the source of the River Aach, which is located in the Rhine Basin. For more information see Mechlem (2003).
25. Many authors criticize this denomination since confinement is a hydraulic state and is not related to the characteristic of being connected or related to surface waters. The correct term would be non-recharging aquifers.
26. The UN Resolution 63/124 is available on the Official Documents System of the United Nations website at: http://daccess-dds-ny.un.org/doc/UNDOC/GEN/N08/478/23/PDF/N0847823.pdf?OpenElement.
27. The full text of the Guarani Aquifer Agreement is available at the Brazilian International Affairs Ministry: http://www.itamaraty.gov.br/sala-de-imprensa/notas-a-imprensa/acordo-sobre-o-aquifero-guarani.
28. Binational Commission for the development of the Bermejo River Basin and Rio Grande de Tarija; Tri-National Commission for the Development of the Pilcomayo River Basin; Binational Commission Bridge Colonia Buenos Aires (COBAICO); Administrative Commission of the Río de la Plata (CARP); Intergovernmental Coordinating Committee Countries of the Plata Basin (CIC); Salto Grande Technical Mixed Commission (CTMSG); and the Administrative Commission of the River Uruguay (CARU).

References

Abu Zeid, K.M. and Elrawady, M.H., 2010. Sustainable development of non-renewable transboundary groundwater: strategic alternatives for the Nubian Sandstone Aquifer System. *In*: UNESCO-IAH-UNEP. *Pre-proceedings of ISARM international conference, "Transboundary aquifers: challenges and new directions"* [CD-ROM]. UNESCO headquarters, Paris, 6–8 December 2010. Paris: UNESCO.

Bakhbakhi, M., 2006. Nubian Sandstone Aquifer System. *In*: S. Foster and D.P. Loucks, eds. *Non-renewable groundwater resources: a guidebook on socially sustainable management for water policy pakers*. IHP-VI. Series on groundwater 10. Paris: UNESCO, 75–81.

Ballesteros, D.A.M., 2009. Examination of the environmental consequences of trade regimes in Latin America. *OPERA*, 9 (1), 173–195.

Borghetti, N., Borghetti, J. R., and Rosa, E.F.F., 2004. *Aquífero Guarani. A verdadeira integração dos países do Mercosul*. Curitiba, Brazil: GIA.

Burchi, S. and Mechlem, K., 2004. *Groundwater in international law. Compilation of treaties and other legal instruments*. Rome: DL Service, UNESCO/Food and Agriculture Organization.

Dellapenna, J.W., 2006. The Berlin Rules on water resources: the new paradigm for international water law. Examining the confluence of environmental and water concerns. *In: Proceedings of the World Environmental and Water Congress*, 21–25 May 2006. Omaha: American Society of Civil Engineers, 1–5.

de los Cobos, G. (2010) The transboundary aquifer of the Geneva region (Switzerland and France): successfully managed for 30 years by the State of Geneva and French border communities. *In*: UNESCO-IAH-UNEP. Pre-proceedings – *ISARM International Conference Transboundary Aquifers: Challenges and new directions* [CD-ROM]. UNESCO, 6–8 December 2010. Paris: UNESCO.

Eckstein, G.E., 2005. Protecting a hidden treasure: the U.N. International Law Commission and international law of transboundary ground water resources. *Sustainable Development Law & Policy*, 5 (1), 5–12.

Eckstein, G. and Eckstein, Y., 2003. A hydrogeological approach to transboundary ground water resources and international law. *American University International Law Review*, 19 (2), 201–258.

Eckstein, Y. and Eckstein, G.E., 2005. Transboundary aquifers: conceptual models for development of international law. *Ground Water*, 3 (5), 679–690.

Giordano, M. and Wolf, A., 2003. Sharing waters: post-Rio international water management. *Natural Resources Forum*, 27 (2), 163–171.

GW MATE (Groundwater Management Advisory Team), 2009. *The Guarani Aquifer initiative. Towards realistic groundwater management in a transboundary context.* The World Bank, GW MATE and Water Partnership Program, Case number 9. Available from: https://protect-us.mimecast.com/s/vVChCo2vOqfXwVJJ2i067OT?domain=urldefense.com www.un-igrac.org/ groundwatermanagement [Accessed 4 August 2010].

Harvey, D., 2003. *The new imperialism.* Oxford: Oxford University Press.

Hochstetler, K., 2003. Fading green? Environmental politics in the MERCOSUR Free Trade Agreement. *Latin American Politics and Society*, 45 (4), 1–33.

Hochstetler, K., 2005. Race to the middle: environmental politics in the MERCOSUR Free Trade Agreement. *In:* P. Dauvergne. *Handbook of global environmental politics.* Cheltenham: Edward Elgar, 348–361.

Laborde, L.C., 2010. The law of transboundary aquifers and the Berlin Rules on water resources (ILA): interpretative complementarity. *In*: UNESCO-IAH-UNEP. *Pre-proceedings of ISARM international conference, "Transboundary aquifers: challenges and new directions"* [CD-ROM]. UNESCO, 6–8 December 2010. Paris: UNESCO.

McCaffrey, S.C., 2001. *The Law of International Watercourses: Non-Navigational Uses.* Oxford: Oxford University Press.

McCaffrey, S.C., 2009. The International Law Commission adopts Draft Articles on Transboundary Aquifers. *The American Journal of International Law*, 3 (2), 272–293.

McIntyre, O., 2010. Fragmentation in international water resources law: reconciling the International Law Commission's 2008 Draft Articles on Transboundary Aquifers with the 1997 UN Watercourses Convention. *In*: UNESCO-IAH-UNEP. *Pre-proceedings of ISARM international conference, "Transboundary aquifers: challenges and new directions"* [CD-ROM]. UNESCO, 6–8 December 2010. Paris: UNESCO.

Mechlem, K., 2003. International groundwater law: towards closing the gaps? *Yearbook of International Environmental Law*, 14, 47–80.

Mechlem, K., 2009. Moving ahead in protecting freshwater resources: the International Law Commission's Draft Articles on Transboundary Aquifers. *Leiden Journal of International Law*, 22 (4), 801–821.

Montano, J., *et al.*, 1998. Acuíferos regionales en América Latina sistema acuífero Guaraní. Uruguay: Universidad Nacional del Litoral Santa Fe.

OAS (Organization of American States), 2005. *Office for sustainable development & environment.* Washington: OAS, Water Project Series, number 7. Available from: http://www.oas.org/dsd/Events/english/Documents/OSDE_7Guarani.pdf [Accessed 7 July 2010]. http://www.oas.org/dsd/Events/english/Documents/OSDE_7Guarani.pdf [Accessed 7 July 2010].

OAS, 2009. *Guarani Aquifer: strategic action program. Acuífero Guaraní: programa estratégico de acción.* Bilingual edition. Brazil: OAS. Available from: http://iwlearn.net/iw-projects/Fsp_112799467571/reports/strategic-action-program/view [Accessed 7 July 2010].

Ribeiro, W.C., 2008. Aqüífero Guarani: gestão compartilhada e soberania. *Estudos avançados*, 22 (64), 227–238. Available from: http://www.scielo.br/scielo.php?script=sci_arttext=S0103-40142008000300014&lng=en&nrm=iso [Accessed 3 January 2011].

Salman, S.M.A., 2007. The Helsinki Rules, the UN Watercourses Convention and the Berlin Rules: perspectives on international water law. *Water Resources Development*, 23 (4), 625–640.

Sappa, G. and Rossi, M., 2010. The North West Sahara Aquifer System: the complex of a strategic transboundary resource. *In*: UNESCO-IAH-UNEP. *Pre-proceedings of ISARM international conference, "Transboundary aquifers: challenges and new directions"* [CD-ROM]. UNESCO, 6–8 December 2010. Paris: UNESCO.

Sindico, F., 2010. The management of the GAS: what role for the emerging international law of transboundary aquifers? *In:* UNESCO-IAH-UNEP. *Pre-proceedings of ISARM international conference, "Transboundary aquifers: challenges and new directions"* [CD-ROM]. UNESCO, 6–8 December 2010. Paris: UNESCO.

Villar, P.C., 2010. Moving toward Guarani Aquifer management: the Brazilian case.*In*: UNESCO-IAH-UNEP. *Pre-proceedings of ISARM international conference, "Transboundary aquifers: challenges and new directions"* [CD-ROM]. UNESCO, 6–8 December 2010. Paris: UNESCO.

Villar, P.C. and Ribeiro, W.C., 2009. Sociedade e gestão do risco: o aquífero Guarani em Ribeirão Preto-SP, Brasil.*Revista geografia Norte Grande,* 43 (2), 51–64. Available from:http://www.scielo.cl/scielo.php?script;=sci_arttext=S0718–34022009000200003=es=iso [Accessed 4 March 2011].

Wolf, A.T., 2007. Shared waters: conflict and cooperation. *Annual Review of Environmental Resources*, 32, 241–269.

Zeitoun, M. and Mirumachi, N., 2008. Transboundary water interaction I: reconsidering conflict and cooperation. *International Environmental Agreements: Politics, Law and Economics*, 8 (4), 297–316.

Keeping it legal: transboundary management challenges facing Brazil and the Guarani

David N. Cassuto and Romulo S.R. Sampaio

This paper examines the legal and ecological problems facing the Guarani Aquifer System. Because the majority of the Guarani Aquifer System underlies Brazil, the Brazilian legal regime forms the paper's principal focus. The importance of the region makes the need for accurate information crucial. Yet relying on such information to manage a complex resource presents risks. Too often, the role of uncertainty in regulating is underplayed. Increasing knowledge over the resource demands categorizing "hard" and "soft" uncertainties, especially those presented by climate change. In addition, regulators must acknowledge the unitary nature of the aquifer while remaining sensitive to differing national priorities.

Overview of regional ecology

The Guarani Aquifer System (GAS) underlies Brazil, Paraguay, Uruguay and Argentina. The GAS is the world's largest aquifer. It was first discovered in 1996 and was named for the indigenous people who have inhabited the region for centuries (Brzezinski 2010). The aquifer contains 30,000 km^3 of water, 1.2 million km^2 of surface area and comprises one of the most important eco-regions in the world (Almanaque Brasil Socioambiental 2008). It is integrally connected through overland rivers (the Parana and the Paraguay) with the Pantanal, the largest wetland in the world, which oversits Bolivia, Paraguay and Brazil (Almanaque Brasil Socioambiental 2008). The GAS recharge zones are primarily located in Brazil and in Paraguay whilst the discharge zones reside in Argentina and Uruguay (Usunoff n.d.). The annual recharge rate is estimated at 45 to 55 km^3 of water, which represents less than 0.2% of the freshwater storage (Schmidt and Larroza 2010). The health of the Guarani and the issues created by that transboundary overlap present a complex management dilemma arising from, among other things, the absence of precise technical and scientific information related to underground water.

Annual extraction levels of the GAS currently hover around 1.04 km^3/year – well within recharge levels of the aquifer as a whole, and approximately .003% of the total (World Bank 2009). At current levels of exploitation, the Guarani Aquifer could offer the four overlying nations fresh water for over 2,000 years (Amore and Tröger 2010).[1] Although impressive at first glance, this statistic should not undermine the importance of integrated management policies among the four overlying nations, especially in light of the exponential increase in pressure on groundwater resources worldwide. Economics plays a significant role in determining how the Guarani will be utilized. Since at present there are significant alternative surface-water sources, it is generally not cost-effective to extract water from the aquifer for irrigation or other heavily consumptive uses. However, as climate change shifts the amount and availability of other water, the economics could change.[2]

Use of the GAS waters varies by region, with the majority going to public water supply, but significant amounts also going toward agricultural and industrial uses and recreation (primarily thermal tourism) (Amore and Tröger 2010).[3] Brazil relies most heavily on the Guarani, extracting 87% of the total water withdrawn (Amore and Tröger 2010). This is due both to the country's size and demographics, as well as the fact that it overlies the majority of the aquifer. The state of São Paulo exerts the greatest pressure (Schmidt and Larroza 2010).[4] The regional ecology as well as socioeconomic complexity present an array of managerial challenges. The transboundary nature of the Guarani Aquifer further complicates management.

Brazilian laws and policies regarding the Guarani Aquifer

We focus here on Brazilian laws and policies regarding the GAS for several reasons. First, Brazilian management practices potentially have the greatest impact on the implementation of the recently signed (but not yet ratified) agreement, the Acordo 2010, which acknowledges the authority of each overlying nation to manage that portion of the aquifer over which it sits. Brazil's actions will have the largest impact on the aquifer's heath and resilience. Its responsibility to implement sound water-management practices is therefore the greatest.

Second, the variety of Brazil's activities on the overlying land requires different and yet integrated management strategies. For example, even within a single activity like agribusiness, water needs and impacts can vary. Intense crop use can lead to considerable drawdown, whereas industrial livestock production in another region might lead to groundwater contamination. The Guarani is thus vulnerable to a wide range of activities and policies some of which, such as land use, do not even necessarily implicate the aquifer directly (Almanaque Brasil Socioambiental 2008). In Brazil, municipalities enjoy considerable legal and managerial authority over zoning, posing their own set of challenges for groundwater management.

Third, from a practical perspective, the background and expertise of the authors permits us far more insight into the international and Brazilian legal framework for transboundary and groundwater management. Nonetheless, we acknowledge that a truly comprehensive analysis of the Guarani must include a similar treatment of Paraguay, Uruguay and Argentina.

The legal and regulatory framework

Historically, Brazilian legislators have paid very little attention to groundwater (Benjamin *et al.* 2005). Consequently, the legal regime dealing with groundwater issues in Brazil is of comparatively recent vintage. Growing pressure over this scarce resource is forcing policymakers to address the regulatory gap between surface water and groundwater. The challenge has been to integrate surface and groundwater management (Benjamin *et al.* 2005).

A new paradigm in water law was established in Brazil by the 1988 Federal Constitution and the 1997 National Water Policy Act (Brazilian National Water Management Policy Act 1997). Prior to 1988, private ownership over water resources was permissible (Pompeu 2006). The 1988 Constitution introduced the notion that the environment is an asset of common use and essential to a healthy quality of life. This principle covers water as well (Freitas 2002). No one owns water and all people shall have equal and unfettered access to it.[5] Codifying statutes on the nature of water as a public good soon followed. For example, Article 99 of the 2002 Brazilian Civil Code states that rivers and oceans are public assets of common use and Article 1 of the National Water Policy Act declares that water lies within the public domain (Brazilian Civil Code 2002).[6] This constitutional and statutory combination places water firmly within the legal category of public assets of common use.

However, the aforementioned shift focused primarily on surface water. The issue of groundwater in the 1988 Constitution was limited to jurisdictional issues regarding the managerial powers of the federal and state governments. Article 20, section III, of the 1988 Constitution entrusts the federal government with managing lakes, rivers and watercourses on lands within its domain, that wash more than one state and that serve as boundaries with other countries (Constituição da República Federativa do Brasil 1988, atr. 20, § III). It also extended jurisdiction over beaches and the territorial sea (Constituição da República Federativa do Brasil 1988, atr. 20, § III). By contrast, Article 26 entrusts States with managing groundwater (Constituição da República Federativa do Brasil 1988, atr. 26). This bifurcated authority creates a serious management problem for aquifers such as the Guarani that underlie multiple states and extend beyond national jurisdiction.

Entrusting individual states with differing priorities and management strategies to manage a resource of multilateral and international significance creates federalist tension and jeopardizes international bilateral agreements. To solve this problem, judges and scholars maintain that the concept of watercourses in Article 20 should be broadly construed to include groundwater that serves as boundaries with other countries and/or wash more than one state (Constituição da República Federativa do Brasil 1988, atr. 20). However, this interpretation is not yet settled and seems to flout the plain language of the Constitution. A constitutional amendment has been proposed to address this issue, but has not yet been ratified by the Congress.

Meanwhile, a set of different regulations has been enacted to close the gap between the groundwater and surface water management regimes. The need for integration has become particularly urgent as a result of ballooning demand for water by agribusiness. In 2001, the National Water Resources Council (CNRH) enacted a series of resolutions aimed at integrating ground and surface-water management (Benjamin *et al.* 2005). In addition, the National Environmental Council (CONAMA) also promulgated groundwater quality standards in 2008 (Sant'Anna 2002).

Those states overlying the Guarani have also taken steps to control access and promote conservation. For instance, the state of São Paulo, the main consumer of water from the Guarani Aquifer, created the State Water Resources Council (CERH-SP) to regulate to protect the State's water resources. It established restricted zones for the perforation of tubular wells in the city of Ribeirão Preto, one of the major consumers within São Paulo, in order to shield the Guarani from contamination. This regulation was approved based on the conclusions of a report produced by the Companhia Ambiental do Estado de São Paulo (CETESB). Another example of a brewing conflict lies in the state of Mato Grosso do Sul, where the passage of Resolution 8 in July 2009 empowers the state Environmental Authority to require permits for property owners seeking to operate wells on their land.

However, as the agency begins rationing permits to prevent groundwater overdraft, property owners whose title predates the 1988 Constitution could potentially file takings claims.

Other policies with the potential to impact indirectly the management and use of waters within the GAS include those related to land use. In addition to the already troublesome jurisdictional problem over groundwater resources, power over land use policies are shared by the federal, state and municipal governments with emphasis on the latter. Article 30, I and II, of the Brazilian Constitution empowers municipalities to legislate over matters of local impact (Constituição da República Federativa do Brasil 1988, atr. 30, §§I–II). The above-stated example of the city of Ribeirao Preto is one such example. Land-use policies must take into account the potential negative impacts on overall management policies for the conservation and sustainable use of the GAS.

Regulating the Guarani Aquifer System

The international legal framework on groundwater is constantly evolving. In general, international laws regarding groundwater build on rules already in place for surface water. Of particular relevance to the Guarani is the 1966 Helsinki Rules on the Uses of the Waters of International Rivers, which was adopted by the International Law Association (ILA) and laid out foundational principles for transboundary water issues (Benjamin *et al.* 2005). This agreement was followed by the 1997 UN Convention on Non-Navigational Uses of Watercourses which led to the 2004 Berlin Rules on Water Resources.

The international legal framework for groundwater also encompasses the 1986 Seoul Rules on International Groundwaters, and the 1994 UN International Law Commission Resolution on Confined Transboundary Groundwater. Further, in 2008 the United Nations General Assembly (UNGA) suggested to member states that they consider adopting the draft articles contained in the Resolution (UNGA Resolution 63/124) (UNGA 2011).

International agreements specifically dealing with the Guarani Aquifer include the 1969 Treaty on the La Plata Basin (Benjamin *et al.* 2005). This treaty provides the foundation upon which the Guarani Aquifer Environmental Protection and Sustainable Development Project was construed. All the aforementioned multilateral agreements laid the foundation for the Acordo 2010. The overlying nations signed the Acordo on 2 August 2010. It will enter into force 30 days after it has been ratified in all four countries. The ratification procedure varies significantly from country to country and may take several years to complete.

Agreement on the Guarani Aquifer (Acordo 2010)

The Acordo 2010 outlines basic principles adopted by all the signatory countries and will represent a step forward for the aquifer's management while still respecting domestic autonomy if ratified by the four nations. However, Article 2 does oblige the signatories to manage the aquifer in accordance with governing principles of international law (Acordo sobre o Aquífero Guarani 2010). This includes the obligation, enumerated in Article 3, to not cause harm to another party or to the environment (Acordo sobre o Acquifero Guarani 2010). It also imposes multilateral obligations which include protecting and conserving the aquifer in a manner assuring multiple, rational, sustainable and equitable uses (Acordo sobre o Aquífero Guarani, Art. 4 2010).[7]

The Acordo will also represent a significant achievement in the field of international water law. To date, only a handful of international groundwater-management agreements exist despite the existence of at least 270 transboundary aquifers, which provide water to millions of people (Benjamin *et al.* 2005).

The agreement adopts a number of important management principles. For example, the four nations agree to share information as well as to inform their fellow signatories of any domestic initiatives that may cause transboundary impact. Furthermore, Article 4 of the agreement acknowledges the multilateral importance of protecting and conserving the aquifer as well as the need to identify areas requiring special attention, especially those near the borders (Acordo sobre o Aquífero Guarani 2010). However, as is often the case in multiparty agreements, the language is broad and may simply be papering over disputed issues (Benjamin *et al.* 2005). Similarly, Article 15, which creates a multilateral commission to oversee and manage cooperation between the parties, does not set out any specific duties or authority of the Commission. Instead, it simply states that the Commission will propound its operating regulations at a later date. This leaves the nascent Commission without a clear mandate.

Concerns over the nature and scope of the sovereignty guaranteed to the signatory nations in Article 2 are already the subject of vigorous debates (McCaffrey and McIntyre 2010). For instance, some would argue that the fact that such language could find itself into a transboundary water agreement in 2010 reflects the enormous difficulty that continues to bedevil international ground and surface-water management (Eckstein 2010). This debate over the extent and understanding of what sovereignty means is unlikely to be resolved any time soon. In the interim, a more practical solution would be to recognize that states can and must collaborate based on a shared vision of existing principles of domestic and international environmental laws.

Because the Guarani – like all transboundary water resources – falls under multiple national jurisdictions and because the policies of the respective nations toward overdraft and pollution may impact the rest of the aquifer, the management strategies of the four countries must be harmonized in order to function effectively. That strategy can derive from principles of reasonable and equitable use (Acordo sobre o Aquífero Guarani 2010, Art. 3, 4) as well as principles of no significant harm (Acordo sobre o Aquífero Guarani 2010, Art. 3, 6, 7). Even as the Acordo recognizes these key precepts, its enforcement mechanisms remain underdeveloped. This lack of a shared approach to enforcement is understandable in light of the different national interests involved. However, its absence could hinder future efforts to manage the resource multilaterally. For example, the Commission that it creates lacks any enforcement power. The parties are merely required to "consult" with the Commission but its recommendations are not binding. (Acordo sobre o Aquífero Guarani 2010, Art. 17). In addition, the Acordo's arbitration clause does not bestow any power or authority not already present in international law (Acordo sobre o Aquífero Guarani 2010, Art. 19). Given the looming challenges of climate change, that reality becomes especially sobering (Hall *et al.* 2008).

Harmonizing domestic policies and building efficient commonly shared management strategies requires a thorough understanding of the four national water-regulatory systems. However, a preliminary and equally important task involves acknowledging the challenges posed by the complexity of each domestic legal groundwater regime. The Acordo offers a broad set of policy objectives. Those objectives only become possible if they do not run afoul of the countries' respective groundwater legal regimes. Consequently, the success of the Acordo depends on the ability of the four countries to harmonize their respective approaches to groundwater management – at least with respect to the Guarani. Opining on the scope of all four countries' legal regimes lies outside the scope of this article. We rather intend to offer some insights into the challenges existing within the Brazilian legal framework, because we believe it crucial that such a framework exist before Brazil can participate effectively in a common and harmonized multilateral regulatory strategy.

Hydrological challenges presented by climate change

The Brazilian legal approach

Climate change presents significant challenges relating to water availability. The situation in Brazil is serious and rapidly worsening. Long periods of drought are becoming more frequent, even in wet states in the south, like Paraná and Rio Grande do Sul. A major diversion project aimed at diverting water from the San Francisco River for the arid northeast region of Brazil offers another example of how water management policies in a country known for its water abundance must now focus on avoiding water shortages.

According to the Intergovernmental Panel on Climate Change (IPCC), there is "*high confidence*" that northeastern Brazil "will suffer a decrease in water resources due to climate change" (IPCC 2007). The report projects significant adverse impacts on agriculture, water supply, energy production and health. While dry regions will become drier, there will also be changes in rainfall patterns and runoff in traditionally humid zones. This will impact water availability and water quality, and will present challenges to infrastructure (Hall *et al.* 2008).

Brazil implemented a national climate-change policy at the end of 2009 that highlights the need to manage natural resources in light of the risks posed by climate change. However, the policy offers guidelines rather than any specific plan of action. Specific regulatory measures were left to the executive branch to enact and also delegated to states and municipalities. In 2010, the first step on regulating the National Climate Change Policy Act (NCCPA) was conceived. Decree n. 7,390 of 9 December 2010 was then enacted with the purpose of regulating the NCCPA. While the Decree provides more specific guidance on the implementation aspects of NCCPA, it is still quite broad and lacks concrete enforceable rules (Presidency of the Republic Casa Civil Cabinet Subcommittee for Legal Affairs 2010).

Incorporating the guidelines of the new climate-change policy act and regulations into existing water law will present significant challenges. The principles embedded in the 1997 National Water Policy Act demonstrate an emerging awareness that water management must adapt to modern environmental realities, including climate change (Hall *et al.*2008). Together with the 2009 National Climate Change Policy Act, principles such as the precautionary approach, intergenerational equity, multiple use and risk assessment now infuse the Brazilian water regime. However, enforcement remains a significant issue for Brazil as well as the other three countries (McAllister 2008).

The Acordo is similarly ill equipped for the challenges of climate change. It contains no reference to climate change nor to existing international climate change law. Given the inevitable impact of climate change on water availability and quality – even on a resource as vast as the GAS – that lack seems short-sighted.

The task now facing the nation involves fashioning an independent regulatory apparatus that can withstand campaigns to manipulate public opinion and undermine sound policy making. Sound water policy making in the post-climate-change world requires acknowledging the principles and guidelines already in place despite the pressures created by large-scale development projects that fail to account for the new water-scarce reality.

Transboundary challenges

The recent evolution of international laws and agreements regarding the Guarani Aquifer offers a first step toward a successful management regime (Benjamin *et al.* 2005). International cooperation enabled the involved countries to assess the challenges imposed by the many ecosystemic and political variables. Those variables range from the threat of climate change to significant and potentially irreversible impacts on different ecosystems including the Pantanal (Hall *et al.* 2008). Inventorying uncertainties is a necessary first step to provide for the needs of the overlying countries (Sunstein 2004). It is also a precondition for a second important initiative: reducing asymmetric information. That, in turn, requires qualified public participation at all levels of policy making.

Development pressures and conservation goals offer fertile ground for conflicting interests within the overlying countries. These conflicts can lead to a tragedy of the commons if not properly managed. On the other hand, overregulation can cause sub-optimal use, thereby impairing development. Optimizing regulation presents an ongoing challenge but the obstacles grow larger when the commons is as complex as the Guarani Aquifer. The complexity and environmental importance of the region, as well as the looming threats presented by climate change, make the need for accurate and detailed scientific and technical information urgent and crucial (Hall *et al.* 2008). Yet relying on such information to manage such a complex natural resource also presents risks (Beck 1986).

Environmental uncertainty can never be eliminated; at best it can be quantified and certain aspects of it reduced (Schroeder 1986). Too often, however, the role of uncertainty in risk assessment and in legal and managerial decisions gets reduced or ignored (Wildavsky 1966). Understanding the role of uncertainty and how to best manage it is necessary to achieve a stable regulatory framework – not just for the Guarani region, but for all transboundary water regimes.

Decreasing uncertainty requires reducing asymmetric information. In other words, policy makers must bridge the gaps among scientists coming from different areas of knowledge as a first attempt to mitigate the impacts of the socio-economic burdens borne by the regulated sectors (Krier 1990). Local and traditional knowledge, particularly for a resource as vast as the Guarani, must also be taken into consideration. Capacity building is crucial. Users must have access to state-of-the-art scientific information couched in language that is accessible to non-specialists. The less asymmetric information, the lower the degree of uncertainty and, consequently, the more likely regulatory decision will be beneficial (Rowe and Frewer 2004). In this context, beneficial means both more legitimate and more procedurally efficient, taking into account the competing uses and needs of different regions within Brazil and within the four overlying nations (Laffont 2000).

With regard to the Guarani, recent developments reveal a laudable collaboration aimed at reducing asymmetric information at all levels: international, regional and local. However, public participation must also play a key role. Vital pieces of the puzzle sometimes are only provided by traditional and local knowledge (Benjamin *et al.* 2005). Another crucial component of a successful multilateral management strategy (and here we refer both to multi-state as well as multinational resource management) lies with coordinating actions and plans. Shared information is an often-overlooked management tool.

Underlying the need for comprehensive information sharing on the Guarani is the need to determine whether management strategies face challenges predicated on hard or soft uncertainty. Soft uncertainty arises "where [a] precise outcome cannot be predicted but a probability distribution can be specified . . . " (Wildavsky 1966). Hard uncertainty occurs "where one does not even know the parameters of the outcomes" (Wildavsky 1966). Identifying which policies fall under which category is crucial.

For example, development issues present a soft uncertainty challenge. Potential threats are, for the most part, knowable and their potential impacts are calculable. Climate change impacts, by contrast, present a problem of hard uncertainty (Posner and Sunstein 2008). The dimensions of the threat are not yet measurable and the impacts remain mostly unknown. In other words, threats posed by development present a known unknown while climate change presents multiple challenges predicated on unknown unknowns.

For the former, local policies organized and coordinated by a common set of principles that acknowledge the transboundary nature of the resource might be most efficient. As mentioned, the recent multilateral agreement exemplifies both the advantages and drawbacks to this approach. For the threats posed by climate change, strategies and policies built at the local level must acknowledge the hard uncertainty underlying the global problem. No local strategy can possibly address the complexity of the dilemma, nor should it. However, local strategies that do not account for global challenges and strategies are doomed to fail. Understanding this reality is particularly important in the case of land-use policies designed and developed under the municipal powers granted by the Brazilian Constitution.

The policy challenges at both the local and international levels will require attention to the precautionary approach, which includes understanding and managing for the catastrophic potential of low-probability events. This involves constructing mitigation policies that navigate dangers presented by unknown and unpredictable events while remaining non-exclusionary and not unduly burdening regulated sectors (Schelling 1992). This task seems impossible. Yet it must be done. The Guarani – and much more – hang in the balance.

Conclusion

In sum, the legal framework for the Guarani represents progress but also highlights the challenges and risks ahead. Brazil's constitutional predicament with respect to transboundary groundwater, as well as other domestic legal hurdles, further complicates an already difficult multilateral management process. Overcoming these challenges and effectively managing the Guarani will require both domestic legal reform as well as increased attention to hard and soft uncertainty questions. Once categorized, policy makers must allow for public participation through the promotion of awareness, capacity building, community involvement and traditional knowledge. In addition, the regulatory framework must acknowledge the unitary nature of the aquifer while still remaining sensitive to differing national and local priorities. This challenge – faced by every transnational water negotiation – looms particularly large with the Guarani because of the size and importance of the resource, and because of its impact on other sensitive and protected regions.

Acknowledgements

The authors would like to acknowledge gratefully the invaluable research work provided by Carolina Gueiros, Brazilian attorney and LL.M. Candidate at Pace Law School, as well as by Shaina Brenner, Stephen Iannacone and Joseph Edgar, J.D. candidates at Pace Law School.

Notes

1. "The static water reserves were calculated as higher than 29,551 km^3 (4,000 km^3 +), but the deep recharge in 1.4 km^3/yr, indicating the occurrence of mining water in confined areas. The exploitable volumes were calculated in 2,014 km^3 (+ 270 km^3), or 6% of SAG reserves, considering the maximum drawdown of water levels by pumping as 400 m. If current exploitation was maintained at 1.04 km^3/yr, through the 1,800 wells that reach the SAG (a total of 8,000 known wells in the region), available reserves could be exploited by more than 2,000 years at current patterns of consumption" (Amore and Troger 2010, p. 2)

2. There are already signs of stress on surface-water supplies. For example, Merin Lake, which is bordered by both Uruguay and Brazil, supplies water for much of Uruguay's rice production. Brazil withdraws significant amounts for irrigation as well; see Ministerio de Vivienda Ordenamiento Territorial y Medio Ambiente (2011). As a result, demands for water from the lake are stressing the regional ecosystem; the most likely candidate for an alternative source is the Guarani.
3. "In general, the main use of the waters of GAS is for public supply (66%). The industrial use (refrigerators, sugar/alcohol plants, etc.) reaches 16%, while in the countryside it is only 5%. Recreational uses in thermal areas have reached 13% and 100% of water in Argentina is intended for tourist use. In Paraguay and Uruguay more than 90% of GAS water is used to supply urban areas" (Amore and Troger 2010, p. 3).
4. "Both in terms of population and groundwater production Brasil has a dominating role in the SAG [Guarani Aquifer System – *Sistema Acuífero Guarani*] region: 87% of the SAG population of 92 million inhabitants live on the Brasilian territory, and about 93% of the recent groundwater production (1,040 million m^3 in 2007) happens in Brasil, mainly in the federal state of São Paulo" (Schmidt and Larroza 2010, p. 2). Besides the state of São Paulo, another seven Brazilian states overlie the Guarani Aquifer: Rio Grande do Sul, Santa Catarina, Paraná, Mato Grosso, Mato Grosso do Sul, Goiás and Minas Gerais.
5. Brazilian law construes equal access according to the principle of isonomy, which allows for differing capacities, needs and uses to dictate water-management and allocation policies.
6. See also Brazilian National Water Policy Management Act (1997, art. 1).
7. It is worth noting that the aforementioned objectives are aligned with the ones found in the 1997 Brazilian National Water Management Policy Act. See Brazilian National Water Management Policy Act (1997, No. 9.433).

References

Acordo sobre o Aquífero Guarani, 2010 [online]. Available from:http://internationalwaterlaw.org/documents/regionaldocs/Guarani_Aquifer_Agreement-Portuguese.pdf [Accessed 19 July 2011].

Almanaque Brasil Socioambiental, 2008. *In*: B. Ricardo and M. Campanili, eds, *Aqüífero Guarani*. São Paulo: ISA, 177–194, 297.

Amore, L. and Tröger, U., 2010. Transboundary Guarani Aquifer System and groundwater management mechanisms. *In: Proceedings of the ISARM2010 international conference "Transboundary aquifers: challenges and new directions"*, 6–8 December 2010. Available from: http://www.isarm.net/publications/325.

Beck, U., 1986. *Risk society: towards a new modernity*. London: Sage, 28–29.

Benjamin, A.H., Marques, C.L., and Tinker, C., 2005. The water giant awakes: an overview of water law in Brazil. *Texas Law Review*, 7 (83), 2218–2243.

Beto, R. and Campanili, M., 2008. *Almanaque Brasil Socioambiental*. Sao Paulo: ISA.

Brazilian Civil Code, 2002. Article 99. Available from: http://www.planalto.gov.br/ccivil_03/Leis/2002/L10406.htm. [Accessed 19 July 2011].

Brazilian National Water Management Policy Act, No. 9.433. 1997 [online]. Available from: http://www.cawater-info.net/bk/water_law/pdf/brazil2001.pdf [Accessed 6 March 2011].

Brzezinski, M., 2010. Regulating transboundary groundwater: big challenges for Brazil. *In*: *Proceedings of the ISARM2010 international conference "Transboundary aquifers: challenges and new directions"*, 6–8 December 2010. Available from: http://www.isarm.net/publications/325.

Constituição da República Federativa do Brasil, 1998. *Presidency of the Republic Casa Civil Cabinet Subcommittee for Legal Affairs*, 2010. Decree no 7,343 [online]. Available from: http://www.planalto.gov.br/ccivil_03/_Ato2007-2010/2010/Decreto/D7390.htm [Accessed 19 July 2011]

Eckstein, G., 2010. *Hydraulic harmony or water whimsy? Guarani Aquifer countries sign agreement* [online]. Available From: http://www.internationalwaterlaw.org/blog/?p=290 [Accessed 5 August 2010].

Freitas, V.P., 2002. Considerações Gerais. *In*: V.P. Freitas, ed. *Águas: aspectos jurídicos e ambientais*. Juruá: Curitiba, 17–18.

Hall, N.D., Stuntz, B.B., and Abrams R. H., 2008. Climate change and freshwater resources. *WTR Natural Resources & Environment*, 22 (3), 34–35.

IPCC, 2007. *Climate change 2007: impacts, adaptation and vulnerability.*

Krier, J.E., 1990. Risk and design. *Journal of Legal Studies*, 19 (2), 787.

Laffont, J.-J., 2000. *Incentives and political economy*. New York: Oxford University Press.

McAllister, L.K., 2008. *Making law matter. Environmental protection & legal institutions in Brazil.* Stanford: Stanford Law Books.

McCaffrey, S.C. and McIntyre, O., 2010. *Sovereignty and cooperative management of shared water resources in a time of shrinking availability: the role of internet international law; fragmentation in international water resources law. Reconciling the International Law Commission's 2008 Draft Articles on Transboundary Aquifers with the 1997 UN Watercourses Convention* [CD-ROM].

Ministerio de Vivienda Ordenamiento Territorial y Medio Ambiente, 2011. Segunda comunicacioun nacional al conferencia de las partes en la convencion: *marco de las nacional unidas sobre el cambio climatico*. Unidad del Cambio Climatico, 226–227(2204) [online]. Available from: http://www.cambioclimatico.gub.uy//index.php?option=com_search&Itemid=5&searchword=Segunda+comunicacion&searchphrase=any&ordering=newest [Accessed 19 July 2011].

Pompeu, C.T., 2006. *Direito de Águas no Brasil*. São Paulo: RT, 41–45.

Posner, E.A. and Sunstein, C.R., 2008. Climate change justice. *Georgetown Law Journal*, 96, 1590.

Rowe, G. and Frewer, L.J., 2004. Evaluating public-participation exercises: a research agenda. *Science, Technology & Human Values*, 29 (4), 520.

Sant'Anna, L.F.H., 2002. General overview of Brazilian environmental law 2002. *SPG International Law Practicum*, 15, 22.

Schelling, T.C., 1992. Some economics of global warming. *American Economic Review*, 82 (1).

Schmidt, G. and Larroza, F., 2010. Pedro Juan Caballero – Ponta Porã. A groundwater transboundary situation between Paraguay and Brasil, 1. *In: Proceedings of the ISARM2010 international conference "Transboundary aquifers challenges and new directions"*, 6–8 December 2010. Available from: http://www.isarm.net/publications/325.

Schroeder, C.H., 1986. Rights against risks. *Columbia Law Review*, 86 (3), 497.

Sunstein, C.R., 2004. *Cost-benefit analysis and the environment*. University of Chicago, Olin Law & Economics Program. Working paper 227.

UNGA, 2008. *UN General Assembly Resolution on the Law of Transboundary Aquifers*. A/RES/63/124. Available from: http://internationalwaterlaw.org/documents/intldocs/UNGA_Resolution_on_Law_of_Transboundary_Aquifers.pdf [Accessed 29 July 2011].

Usunoff, E., n.d. *Web based information for integrated water resources management of a multi-national aquifer: the global environment facility project on the Guararí Aquifer* [online]. Instituto de Hidrología de Llanuras. Available from: http://www.waterweb.org/wis3/presentations/30_Usunoff_paper.pdf [Accessed 19 July 2011].

Wildavsky, A., 1966. The political economy of efficiency: cost-benefit analysis, systems analysis, and program budgeting. *Public Administration Review*, 26 (4), 296.

World Bank, 2009. *Implementation completion and results report* (Project ID: Po68121 L/C/TF Number(s): TF-50950 TF-51223. Report No.: ICR00001198. Environmental Protection and Sustainable Development of the Guarani Aquifer System Project [online]. Available from: Http://www-wds.worldbank.org/external/default/WDSContentServer/WDSP/IB/2009/08/13/000333037_20090813231853/Rendered/PDF/ICR11980P068121ICOdisclosed08112191.pdf [Accessed 19 July 2011].

Sustainable transboundary groundwater management under shifting political scenarios: the Ceylanpinar Aquifer and Turkey–Syria relations

Mehmet Öztan and Mark Axelrod

Recent scholarship evaluates how peace and conflict affect resource sustainability. The authors of this article address how enhanced Turkey–Syria cooperation could impact management of the Ceylanpinar Aquifer, which flows beneath both countries and accounts for over10% of Turkey's transboundary groundwater discharge. Although the Euphrates–Tigris river system apportionment has been critical for bilateral relations, allocation of the Ceylanpinar Aquifer has been neglected. The authors develop a simplified conceptual model to simulate water flow in the aquifer. Taking this model as a reference point, this article explores how political changes may impact aquifer sustainability, and proposes data-integrated modelling for sustainable management.

Introduction

Upstream Turkey and downstream Syria have long engaged in politically driven confrontations over shared water resources. In recent years, the political relationship – though still facing significant challenges – has improved overall, leading to possibilities for enhanced water cooperation between the two countries.

This paper analyzes likely impacts of interstate political cooperation on the shared groundwater resources of the Ceylanpinar Aquifer. The Ceylanpinar is a transboundary aquifer with a recharge zone located in southeastern Turkey that discharges through the Ras al-Ain Springs in northern Syria.

Following a summary of existing scholarship on conflict, peace and sustainability, we discuss recent developments in the region. We then introduce a conceptual model to simulate groundwater flow in the Ceylanpinar Aquifer based on limited data available from Turkish sources. Although it is a simplified representation of the aquifer, the model provides an opportunity to address major concerns regarding sustainable use of the aquifer. In particular, we employ the model to propose a data-integrated modelling framework based on past experience and future expectations of the Turkey–Syria political relationship.

Conflicts and water use

There is an extensive literature regarding the impacts of war on environmental degradation. Reuveny *et al.* (2010) find that warfare increases the overall level of environmental stress, particularly for developing countries fighting on home territory. As a result, we might expect that improved relations could lead to greater environmental protection.

On the other hand, conflict may reduce economic activities that lead to resource strains, exemplified by a decline in fishing during the Second World War and reduced hunting following the 1991 Gulf War (Reuveny *et al.* 2010, p. 750). Militarized buffer zones are de facto protected areas, especially when mines threaten injury to those seeking access (Carpenter 1992). A peaceful border and the removal of unexploded ordinance may therefore increase strain on natural resources.

Border-mine removal may therefore open up nearby lands to increased agricultural production and therefore additional water usage. In addition, increased Turkish–Syrian cooperation may involve further development of water resources as in the example of Orontes River Dam noted below. On the other hand, enhanced cooperation may also include collaborative sustainable management activities. Our initial simulation, therefore, examines how mine removal is likely to affect future sustainable use of Ceylanpinar waters. We will conclude by outlining other scenarios for subsequent analysis.

Turkey proposed to begin Euphrates management cooperation with "inventory studies of water resources" (Turkish Ministry of Foreign Affairs 1996, p. 12) that would eventually expand to "a simulation . . . to analyze water demand and supply balance" (ibid., p. 13). However, little information on groundwater has been available to support comparable efforts for transboundary aquifers. While abundant information exists above ground, a lack of groundwater data has limited the ability to determine likely effects of different management practices. As a result, international cooperation over sustainable resource use has also been constrained (Jarvis *et al.* 2005).

In response to these concerns, we employ a conceptual model that allows us to better understand the groundwater dynamics in the face of changing resource use. We identify the likely impact of different management scenarios, in hopes that this information may support careful management decisions by Turkish and Syrian leaders eager to promote long-term cooperation, rather than merely short-term economic development, along the border.

Turkish–Syrian water relations

Turkish industrial and agricultural water demands grew significantly from the 1950s to 1990s, with dam building increasingly employed as a strategy to meet these demands. Only nine dams were built prior to 1960, but 172 were in place as of 1997 (Turan 2003, pp. 191–193). With this increased control capacity, Turkey has occasionally threatened to limit or completely shut off the flow of surface water to Syria. In 1990, Turkey stopped Euphrates waters from entering Syria, ostensibly to fill the new Atatürk Dam. Although the action was intended for technical purposes, Syrian and Iraqi authorities objected that this action limited the supply of water for downstream agricultural users (Starr 1991, pp. 29–31). The Turkish government threatened to cut off water flow again later the same year, and other countries also pushed Turkey to employ the strategy as a punishment for Iraq's invasion of Kuwait (Gleick 1993, pp. 88–89). Although Turkey has not subsequently used this strategic tool, the threat remains plausible and severe.

In addition to the threat of strategic water cutoffs, the ongoing Southeast Anatolia Development Project (known by its Turkish acronym, GAP) could potentially reduce the flow of Euphrates waters to Syria. GAP is designed to increase agricultural production and employment in southeastern Anatolia (Turan 2003, pp. 198–199), with the related goal of standardizing produce quality for Turkish participation in foreign markets (Turan 2003, p. 195). Concerns have been raised that GAP dams could reduce the flow of Euphrates waters to Syria by up to 40%, or even 50%, at full operation (Gleick 1993, p. 89, Amos 2010).

Despite these highly publicized concerns about Euphrates waters, very little concern has been raised about shared groundwater resources. We next turn to current groundwater conditions in both countries to demonstrate that similar concerns exist with regard to this shared resource.

Groundwater use in Syria

As of 2010, agriculture contributed about 18% to Syria's gross domestic product (GDP), and employed nearly 17% of the country's workforce. Intensive agricultural use, industrial expansion and population growth of 1.95% have increased pressure on water supplies (Central Intelligence Agency 2011). Total irrigated area increased from 650,000 ha in 1985 to 1.3 million ha (of which 63% is located in the Euphrates–Tigris Watershed) in 2002, as a result of the Syrian government's policy to maintain food security. This expansion was dependent upon groundwater-irrigated area increasing to 764,000 ha in 2002, almost 60% of the country's total irrigated area (Salman and Mualla 2003).

Expanded groundwater use for irrigation has resulted in water deficits and over-exploitation of groundwater in most watersheds across the country. As of 2002, the total number of pumping wells in the country was above 170,000. More than 55% of these sources were unlicensed. At that time, the continuous decline in groundwater levels was reported to affect spring flows across the country. In the Khabour Watershed where the Khabour River receives most of its water resources from the Ras al-Ain Springs, overuse of groundwater was considered a serious problem (Salman and Mualla 2003). In 2000, it was reported that, together with the adverse effects of the dry growing seasons, the downstream flow of the Khabour River was drastically reduced by approximately 2,000 unlicensed pumping wells drilled for irrigation in the Ras al-Ain area (Assyrian International News Agency 2000). There is no indication of any decrease in pumping since that time.

Kattan (2001) reports annual discharge values of the Khabour River for the period from 1942 to 1994. According to his analysis based on unpublished data from the Syrian Ministry of Irrigation, there was a significant reduction (from 42 m^3/s to approximately 22 m^3/s in 1993) in the flow of the Khabour River (that is, the discharge of the Ras al-Ain Springs) after the late 1980s. This decline is consistent with government policy expanding nationwide irrigation after 1985, and the marked increase in groundwater use for irrigation noted above. This shift indicates that regional management interventions may be necessary to address the river's flow.

As a result of a recent policy change, the Syrian government intends to inspect all illegal wells in the country to regulate pumping discharges. This monitoring would lead to decisions on whether to grant a licence for a well with a groundwater extraction plan or to close it. However, questions remain about the capacity to enforce this policy (Kibaroglu *et al.* 2006).

Groundwater use in Turkey

As of 2005, Turkey irrigated an area of 27,000 ha with groundwater from the Ceylanpinar Aquifer. The irrigation component of the GAP aims to increase groundwater-irrigated area in the Ceylanpinar and neighbouring watersheds (Kibaroglu *et al.* 2006). TUMAS (1991) concludes that in the long term, based on water demand for selected crop patterns, groundwater resources of the Ceylanpinar Aquifer could potentially irrigate an area of 150,000 ha.

In the Harran Watershed to the west of Ceylanpinar, 400 of 1,700 pumping wells were identified as unlicensed, with the amount of total water extraction by these wells unknown. It is likely that additional unlicensed and uninspected wells are also used for irrigation in the Ceylanpinar Watershed. A 2000 World Bank study stressed that discharge of the Ras al-Ain Springs has declined due to rapid extraction in both countries (Kibaroglu *et al.* 2006).

Shared water supplies such as in the Ceylanpinar Aquifer offer an opportunity for further conflict or cooperation. Unlike river flows, groundwater cutoffs have not been used as a strategic threat in this relationship. Due to the aquifer's contentious border location, however, the overlying lands also have not been developed for agricultural production. In the 1950s, for security reasons, Turkey planted more than half a million land mines along the 510 km of the national border shared with Syria, in a strip with an average width of 500 m (International Campaign to Ban Landmines 2001, Hurriyet Daily News 2009). As a result, Turkish use of Ceylanpinar waters has been limited.

Recent developments

Turkey has in recent years become a more important player in the northern Middle East and has attempted to engage Arab neighbours in ongoing collaborative efforts (Fuller 2008, pp. 93–96, Traub 2011). As tensions subside, Turkey has engaged in conversations with Syria and Iraq about water cooperation (Watson and Comert 2009). The Turkish government has also sought the promotion of development projects in neighbouring countries such as Syria, including a new dam on the Orontes River flowing from Syria into Turkey (Hurriyet Daily News 2011).

As these types of cooperation continue and expand, we anticipate that water usage will also shift. For instance, in 2009, the Turkish government passed a law to sweep the aforementioned mines along the entire border (Hurriyet Daily News 2009). After the mines are removed, approximately 25,000 ha of additional agricultural area may be expected. The soil in this region has potential for organic agriculture since chemical fertilizers or pesticides have not been applied for more than five decades.

Characterization of the study area

Location and physical characteristics

Exact dimensions of the Ceylanpinar Aquifer are not known. However, the upstream portion of the aquifer is primarily located within the Ceylanpinar Watershed, part of the Euphrates–Tigris river system. The watershed is located in the southeastern Turkish provinces of Sanliurfa and Mardin, and covers approximately 7,700 km^2. The aquifer's downstream portion extends beyond the national territory of Turkey, towards the Lower Balikh and Lower Khabour Watershed in northern Syria (Figure 1) (Kibaroglu *et al.* 2006).

Hydrogeologic characteristics

The Ceylanpinar Aquifer mainly consists of limestone geologic units from the Ecocene (Oldac *et al.* 1971), and has the greatest groundwater reserve among aquifers in Turkey.

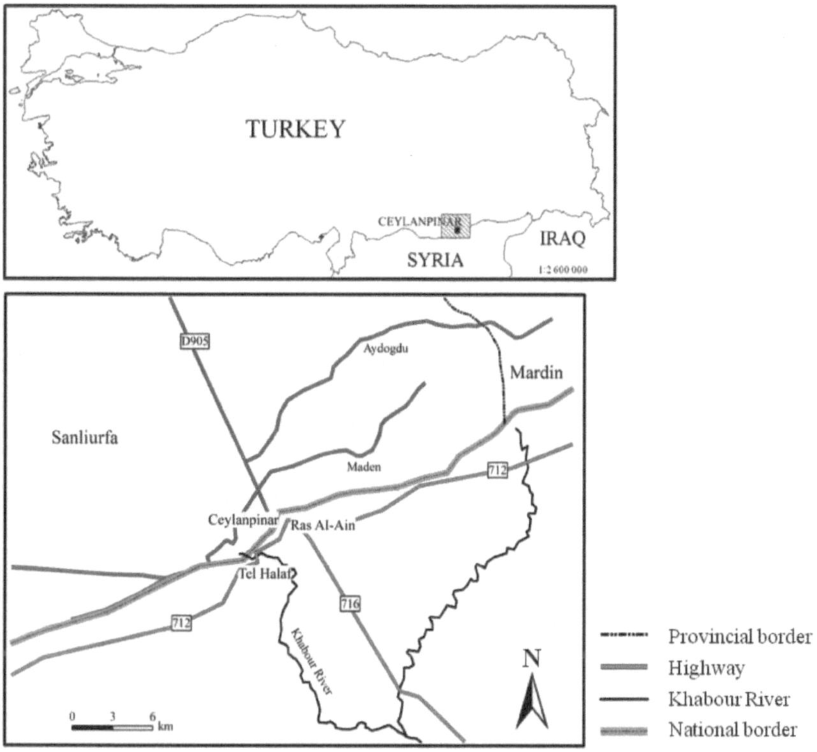

Figure 1. Location of the study area.

The aquifer is recharged in Turkey, and discharges into the Ras al-Ain karstic springs located in northern Syria (Figure 1). The portion under Turkish territory is mostly unconfined and hydraulically connected to the Khabour River, the largest perennial tributary to the Euphrates River in Syrian territory. While the Khabour River originates in Turkey, close to the national border with Syria, its main source water comes from the discharge of the Ras al-Ain Springs. At the location of the springs, groundwater levels intersect the land surface. There are a total of 18 springs (five in southern Turkey, and 13 in northern Syria), with a total discharge rate of 43 m^3/s (1 m^3/s of discharge in Turkey, 42 m^3/s of discharge in Syria) (Burdon and Safadi 1963, Oldac *et al.* 1971). The springs, therefore, account for more than 11% of Turkey's total annual groundwater discharge to neighbouring countries (FAO 2010). The springs are enclosed within a relatively small 1,200 m radius, where low regional precipitation prevents the diffusion of groundwater flow over a large area. As a result, the Ras al-Ain Springs are considered some of the greatest limestone springs on earth (Burdon and Safadi 1963).

Conceptual model of the Ceylanpinar Aquifer

In this section, we present a conceptual model of groundwater flow in the Ceylanpinar Aquifer. Data for the model are obtained from the Ceylanpinar Plain Hydrogeological Investigation Report prepared by Turkey's General Directorate of State Hydraulic Works (known by its Turkish acronym, DSI) in 1971, and the Ceylanpinar Region Groundwater Feasibility Study Report prepared by the Turkish Engineering, Consulting and Contracting

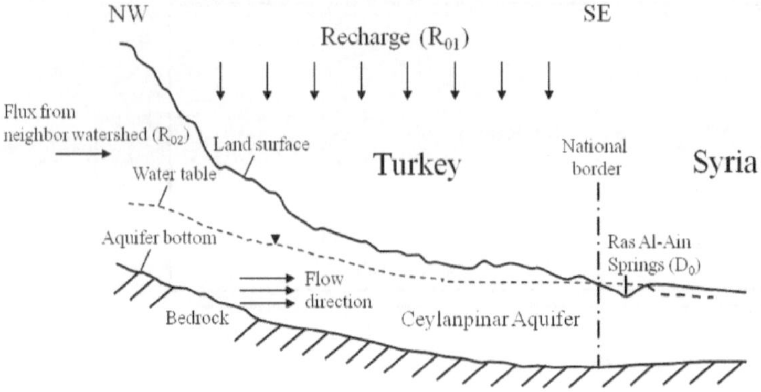

Figure 2. Conceptual illustration of the Ceylanpinar Aquifer before pumping.

Co. (TUMAS), a partially state-owned company, in 1991.[1] Major features of the conceptual model are illustrated in Figure 2.

Groundwater flow in the aquifer is represented with Equation (1):

$$\frac{\partial}{\partial x}\left(K_x(h_C - e)\frac{\partial h_C}{\partial x}\right) + \frac{\partial}{\partial y}\left(K_y(h_C - e)\frac{\partial h_C}{\partial y}\right) + q_r - q_d = 0. \tag{1}$$

where h_c is the groundwater level, K_x is hydraulic conductivity in the x direction, K_y is hydraulic conductivity in the y direction, e is bottom elevation, q_r is the source term (that is, recharge), and q_d is the sink term (that is, discharge through springs, and pumping where applicable).

The Ceylanpinar is an "evolved karstic aquifer". Towards the discharge area, there are underground galleries that hold groundwater (Oldac *et al.* 1971). Based on this evidence, it is assumed that the groundwater flow velocity stays within the limits dictated by the laminar flow conditions, and Darcy's Law is applicable for the conceptual model. The whole aquifer medium is assumed to be continuous, and the Dupuit-Forccheimer assumptions are also assumed to be valid for representing the whole regional groundwater system with one conceptual layer. The hydraulic conductivity field is also assumed to be isotropic, in other words, groundwater flows in the direction of the steepest hydraulic gradient; for further explanation of these concepts and assumptions, see Domenico and Schwartz (1998) and Anderson and Woessner (2002).

The Ceylanpinar Watershed boundaries comprise the model's borders. These boundaries delineate the drainage area of the streams within the watershed, except for the southern boundary, which is the political border between Turkey and Syria. The watershed boundaries are assumed to coincide with the groundwater-basin boundaries, and there is no lateral flux entering/leaving the groundwater system along the eastern, western and southern model boundaries. The base of the aquifer is assumed to be impermeable, and there is no vertical flux entering or leaving the system through the base.

The land surface elevation is the upper bound of the aquifer, and is conceptualized as a seepage surface. Most of the streams within the watershed are seasonal (Oldac *et al.* 1971). Because of their limited contribution, they are not included in the conceptual model.

Recharge into the aquifer comes from two main sources: (1) precipitation and surface runoff; and (2) subsurface flux from the neighbouring watersheds to the west (Harran Watershed) and east (Mardin Watershed) (Oldac *et al.* 1971, TUMAS 1991). TUMAS

(1991) provides a two-dimensional water-table map that was manually drawn based on the measured water levels in aquifer test wells. Even though this map shows the groundwater interaction between the surrounding watersheds and the Ceylanpinar Watershed, the amount and distribution of water entering the system through the hydrologic boundaries on the west and east is unknown. Furthermore, the TUMAS (1991) map clearly indicates that there is a flux contribution into the Ceylanpinar Watershed through the northern hydrologic boundary on the high topography zone of the watershed. The northern boundary is therefore conceptualized as the constant flux boundary that contributes subsurface flow into the watershed. The Ras al-Ain Springs are conceptualized as a drain that discharges the groundwater from the aquifer. Even though these springs are located in northern Syria, they are close to the town of Ceylanpinar in Turkey, and in the conceptual model they are located very close to the national border.

Numerical model

A finite-difference numerical model is created based on the conceptual model described above. The numerical model uses Interactive Groundwater® software as the groundwater flow simulator (Li and Liu 2006). The Ceylanpinar Aquifer is represented with one computational layer in the model. The model simulates the steady-state flow conditions. The number of grids in x and y directions are 125 and 129, respectively, where the grid size in both directions is approximately 860 m. The resolution of the data used in model conceptualization does not offer detailed spatial parameter definitions and the grid size is sufficiently fine to simulate the regional-scale groundwater flow in the aquifer. TUMAS (1991) provides data from more than 180 water wells, including well locations, water levels, land-surface elevations and the aquifer's bottom elevations recorded in test wells. These data are geo-referenced through digital files to assign the aquifer's upper and lower elevations at the well locations. The data are then interpolated within the model domain (that is, Ceylanpinar Watershed) to generate the spatial field of the aquifer's upper and lower elevation.

The water budget in Oldac *et al.* (1971) is used to assign the amount of water that enters the aquifer in the models: (1) recharge from precipitation and surface runoff (R_{01}) is added together into the aquifer and is approximately 9 m^3/s; and (2) subsurface flux (R_{02}) that enters the aquifer through the constant flux boundary is the difference between the total discharge of the Ras al-Ain Springs (43 m^3/s) and R_{01}, and is therefore equal to 34 m^3/s. Based on the information provided by Burdon and Safadi (1963), Oldac *et al.* (1971) and TUMAS (1991), 348 m (above sea level) is assigned as the average altitude of the springs.

Resolution of the aquifer test data available in TUMAS (1991) is not fine enough to generate the regional-scale hydraulic conductivity distribution of the aquifer. The model domain is therefore divided into multiple effective hydraulic conductivity zones based on the watershed geology and water-table map provided with TUMAS (1991). Then the model is iterated, through manual calibration, until an optimum set of conductivity values provides a good match between simulated water levels and observed pre-pumping levels measured in wells.

Final results of the model calibration indicate that the hydraulic conductivity of the aquifer has a tendency to increase towards the south, approaching the Ras al-Ain Springs. Final calibrated hydraulic conductivity value for this area is 1,000 m/d, a finding that is consistent with that of Oldac *et al.* (1971): "Underground karstic galleries that hold ground water exist in the southern portion of the watershed."

Simulated groundwater levels are shown in Figure 3. This map indicates that groundwater flows from northwest (NW) to southeast (SE) of the watershed (that is, towards the Ras al-Ain Springs), in the direction of the regional-scale hydraulic gradient, which decreases in the southern portion of the watershed. Comparison between simulated and measured water levels at the well locations is shown in Figure 4.

Two model statistics are used to measure the model's goodness-of-fit to the data: normalized standard deviation of the residuals (NSDR) and the coefficient of determination

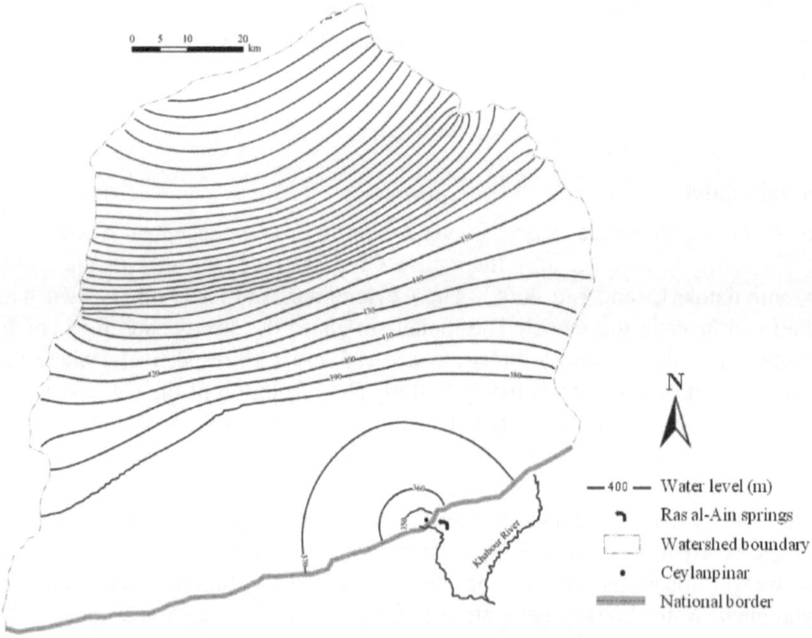

Figure 3. Simulated groundwater levels in the Ceylanpinar Watershed.

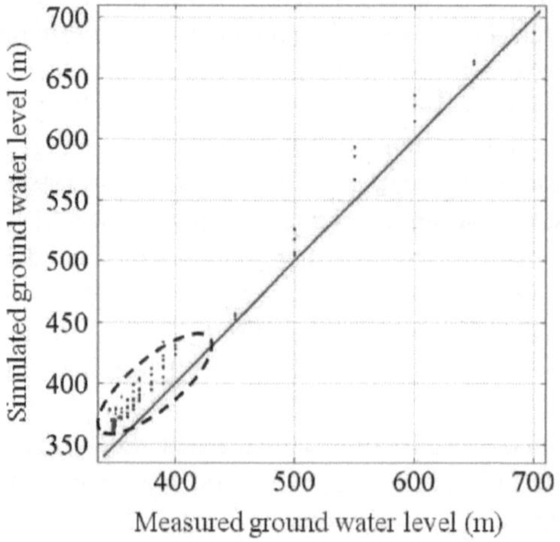

Figure 4. Comparison between the simulated and measured groundwater levels.

(R^2). NSDR and R^2 are 0.02 and 0.99, respectively. For regional-scale groundwater modelling purposes, NSDR values below 0.10 are considered sufficient to provide a good fit to data (Kuniansky *et al.* 2004).

Even though the model provides a reasonably good approximation of water levels for the pre-pumping conditions of the aquifer, it is important to emphasize some potential reasons for the bias that is noticeable especially at lower elevations (circled in Figure 4) close to the Ras al-Ain Springs: (1) grid size is not fine enough to capture local-scale water level changes, and other data uncertainties remain as well; (2) one computational layer can only represent the average water levels, rather than demonstrating them in detail; and (3), most importantly, the model is based on a water budget developed in 1971 and compared with observed water levels that were measured 20 years later. It is likely that the increase in groundwater use for irrigation after 1985 in Syria, and/or groundwater extraction after 1971 in Turkey, decreased the measured water levels along the border (circled in Figure 4). Since the model does not simulate the pumping wells, drawdown caused by newer wells would not be incorporated. The more recent data, therefore, should be expected to come in at a lower level.

Sustainable management of the Ceylanpinar Aquifer

Cooperation on joint resource development and the obligation not to harm other states together form the concept of sustainable transboundary-aquifer management (Puri *et al.* 2001, Kibaroglu *et al.* 2006). In other words, upstream and downstream states both should be responsible for maintaining sufficient groundwater flow for regional development projects, without adversely affecting groundwater levels.

Long-term sustainable groundwater development through pumping requires the change in water storage in an aquifer to be zero (Equation [2]) (Bredehoeft 2002).

$$\Delta R_0 - \Delta D_0 = P \tag{2}$$

where ΔR_0 is change in virgin-groundwater recharge rate, ΔD_0 is change in virgin-groundwater discharge rate, and P is pumping.

To reach a sustainable state, pumping (P) must be balanced with a change in ΔR_0, and/or a change in ΔD_0, and must not exceed $\Delta R_0 - \Delta D_0$, or capture. If we assume that the groundwater-recharge rate (initially $R_{01}+R_{02}$ for the conceptual model developed here) is only a function of natural conditions (for example, rainfall, vegetation, landscape slope, soil permeability), is unaffected by pumping, and initial recharge does not change, then ΔR_0 should equal zero (Bredehoeft 2002, Devlin and Sophocleous 2005). Therefore, even without quantitative knowledge of recharge rates, sustainable pumping can be determined based on the change in the virgin-groundwater discharge rate, ΔD_0, of the Ras al-Ain Springs (Equation [3]) (Devlin and Sophocleous 2005).

$$P = -\Delta D_0 \tag{3}$$

Sustainable groundwater management can be possible if the extraction (that is, pumping) in both Turkey and Syria stays within the limits for which the aquifer's groundwater storage is not adversely affected. As Figure 5 indicates, over-pumping of groundwater in the region may reduce water levels, eventually causing the springs and Khabour River to run dry

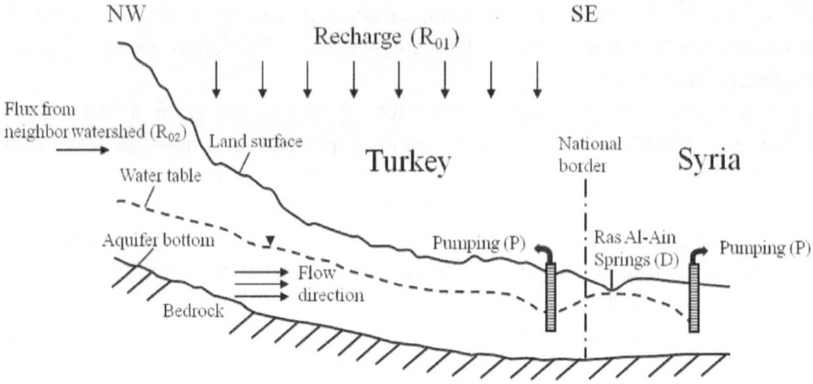

Figure 5. Conceptual illustration of non-sustainable pumping of the Ceylanpinar Aquifer.

(compare water table in Figures 2 and 5). If overuse continues, it may not be possible to recover the aquifer's resources within a reasonable timeframe.

In order to demonstrate the impact of increased agricultural activity along the border, we simulate the impact of additional pumping in this area. Based on the Turkish government's mine removal policy discussed above, we anticipate new farming in the 500 m band along the border. The Ceylanpinar Watershed runs along an approximately 80 km stretch of the national border, meaning a newly farmed area of 40 km^2 (or 4,000 ha) in the watershed. If this area were planted with organic cotton, which requires approximately 1,300 mm of water over the growing period in hot arid climates (Critchley *et al.* 1991), then a total of 3.4 m^3/s of water would be needed to irrigate cotton in this area for each growing period of approximately 180 days (Critchley *et al.* 1991). We simulate the groundwater use for cotton through 20 pumping wells along the national border to anticipate the impact of pumping on groundwater levels in the region (Figure 6). Figure 6 clearly shows that even such minimal added pumping would affect groundwater levels, particularly in the southern portion of the watershed, and impose drawdown on the water levels. If additional projects are introduced without careful management interventions, we anticipate the possibility of an eventual shift to unsustainable water use (for example, dry springs at Ras al-Ain). These results suggest initial support for international relations theories that link peace to increased resource use.

However, as Turkey and Syria open lines of communication, they may also benefit from efforts to engage in joint management activities. For example, Turkey may mitigate increased agricultural use by building technical or enforcement capacity among Syrian counterparts to reduce drawdown on the Syrian side.

Conclusion: data-integrated groundwater modelling for sustainable management

Groundwater is often referred to as a hidden resource (Heijde *et al.* 1985) that cannot be directly observed. The observation of this resource is more difficult compared to surface-water resources, because of: (1) spatial heterogeneities; (2) complex geology and topography-driven flow dynamics and boundary conditions; and (3) coupled interactions with surface-water bodies in the same hydrologic system. Changes in water levels, quality and fluxes due to human-induced stresses and management plans further complicate the process. Relaxation of these complexities and proper conceptualization of an aquifer for groundwater development are closely linked to availability of field data and mathematical

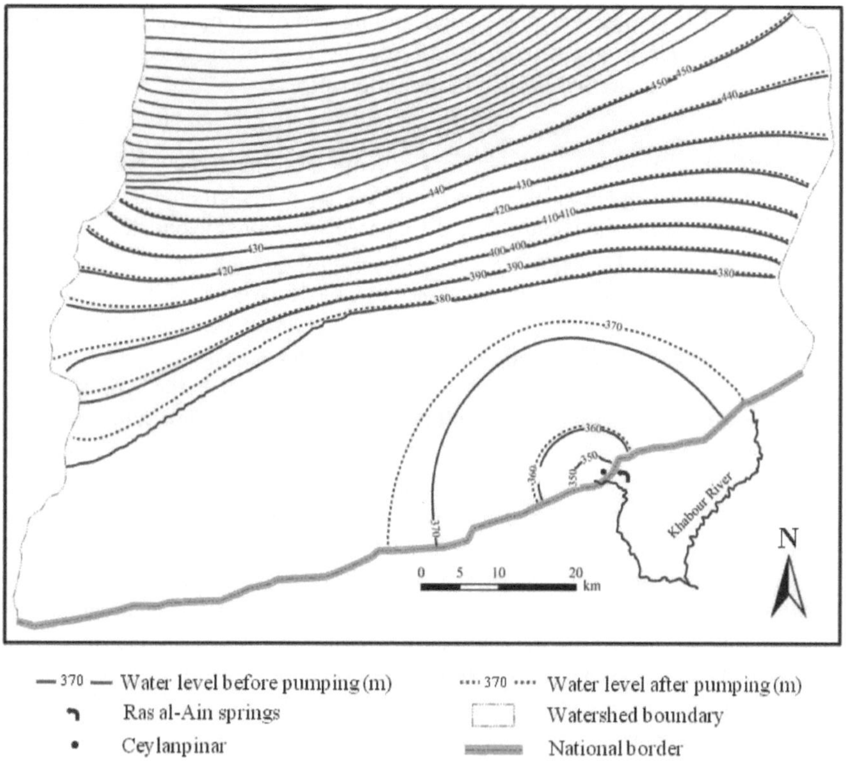

Figure 6. Two-dimensional comparison between simulated groundwater levels before and after irrigational pumping due to the removal of mines along the national border.

models that can transfer those data into information on the groundwater systems (Moore 1983, Heijde *et al.* 1985, Bear *et al.* 1992, Watkins *et al.* 1996, Gogu *et al.* 2001, Mende *et al.* 2007).

Data requirements

While Turkey and Syria fully benefit from the exploitation of the Ceylanpinar Aquifer, poor planning and management of aquifer resources could lead to inefficient use and depletion. The cost of such management failures would be borne by both parties (Chermark *et al.* 2005, Jarvis *et al.* 2005).

While current data can be used to simulate the pre-development aquifer dynamics and the impacts of basic management scenarios, many assumptions are required. As Dimitrov (2003) has noted, cooperative resource management is unlikely without clear understanding of cross-border consequences. Additional data from Turkey and Syria, and accurate scenario modelling, therefore, could facilitate further cooperation between these governments, based on management decisions that involve complex cost–benefit trade-offs. Within this context, developing an effective and sustainable groundwater-management plan for the Ceylanpinar Aquifer requires a cooperative data-exchange programme to be conducted by Turkey and Syria to monitor the aquifer's hydrologic/hydrogeologic parameters, much as Turkey has proposed for the surface-water system. Such a cooperative programme would provide the data essential for generating a quantitative perspective on the aquifer's status and for validating the conceptual understanding of the groundwater system (Puri and Naser 2003), thereby clarifying the interests affected by joint management activities. These essential data include but are not limited to the following:

(1) Additional data from Syrian territory is required to accurately conceptualize the aquifer, and to quantify the likely impacts of management practices. Clearly, Ceylanpinar Watershed boundaries do not precisely coincide with the groundwater-basin boundaries. Furthermore, the national border is not a hydrologic boundary. Extension of the aquifer's boundary in Syrian territory is therefore critical for identifying the aquifer's domain, thus reducing the number of modelling assumptions and improving the understanding of aquifer dynamics.
(2) Variations in discharge of the Ras al-Ain Springs, as well as the hydraulic connection between the Khabour River and groundwater in the region, and monitoring of river levels are needed to inspect the impacts of groundwater use on the system's integrity, and to maintain sustainable pumping in the aquifer. Previous observations (Kattan 2001) indicate that there is a strong correlation between groundwater irrigation in Syria and the spring discharges. These data are especially important to identify the impacts of management activities in Syrian territory.
(3) More investigation wells in the upstream portion of the Ceylanpinar Watershed and along both sides of the national border are needed for a better representation of the aquifer's upper and lower elevations, and current groundwater levels. The measured groundwater levels not only provide critical information for model calibration, but when continuously observed they also provide an opportunity to determine the impacts of groundwater use in the region and in the vicinity of the Ras al-Ain Springs. Investigation wells also help to identify the subsurface extensions of different geologic units within the aquifer and their ability to conduct water.
(4) Amount of precipitation, land use, evaporation and soil characteristics are some of the critical parameters that are needed to calculate the recharge that enters the aquifer. Monitoring the spatial and temporal variations of these parameters on both sides of the border, and therefore of aquifer recharge relative to discharge, is important for understanding the groundwater system's dynamics. In addition, future modelling should account for expected regional impacts of anthropogenic climate change that would alter both precipitation and temperature conditions.
(5) Perhaps the most important data that need to be collected are the number and location of both licensed and unlicensed pumping wells, as well as the extraction rate of these pumping wells on both sides of the border. These data are needed to interpret the impacts of human-induced stresses on the aquifer's resources. Our simulation is based on data from a sparsely used hydrologic system, but there is increasing evidence that baseline conditions have changed since 1971, potentially making the system even more fragile than our model suggests.

The conceptual model presented above utilizes hydrologic data provided by Oldac *et al.* (1971) to simulate the groundwater flow prior to distinguishable agricultural development. If Turkey and Syria exchange existing groundwater-related data that date back to the period during which groundwater use was insignificant, it may become possible to simulate the initial conditions (that is, water levels and water budget) of the aquifer more accurately. Once natural conditions are known, pumping can be imposed to simulate the drawdown, and variations in discharge with respect to agricultural-development periods. Once all historical data are available to construct a groundwater database, it becomes significantly easier to append new data as they become available for conceptual and numerical modelling. As a result, governments can predict the response of the groundwater system to contemplated management plans before they are implemented (Bear *et al.* 1992).

Management scenarios

As we demonstrate above, mine removal without careful management decisions promoting sustainable use could initiate long-term problems for water management in the Ceylanpinar Aquifer. However, the modelling framework we propose would also allow managers to determine what land-use practices could be sustainable if and when this region becomes useful for organic agriculture. Based on the selected crop patterns, efficiency of irrigation systems and economic incentives provided to farmers, groundwater needs can be determined through a data-integrated modelling framework. With the inclusion of these various management factors, local authorities can determine the most efficient long-term management practices for this important body of water.

In addition to the removal of mines along the Turkey–Syria border, many other scenarios can be contemplated as these countries move towards further cooperation. For instance, we can envision how farmers would react to increased prices for organic produce, and therefore how they can be provided incentives to choose production that is both profitable and sustainable for the long term. We would also have the capacity to evaluate whether different crop patterns would be appropriate in different seasons, as discharge rises and declines in wet and dry seasons, respectively. As noted earlier, robust models would also allow us to extrapolate over the long term by incorporating the impact of anthropogenic climate change on water availability in the region.

Other scenarios would also provide essential information for decision making. For instance, another possible result of enhanced cross-border cooperation would be increased Turkish support for agricultural activities in northern Syria. In southeast Anatolia, the Turkish government intended to use water development (GAP projects in particular) to provide economic opportunities for the alienated Kurdish population. While some analysts question the effectiveness of this strategy (Olson 1996), there are historical precedents for economic development projects as part of peace agreements elsewhere in the world. Economic support and shared concerns have, in many situations, enhanced former rivals' reliance upon each other, as well as long-term incentives for peaceful cooperation (Keohane and Nye 2001). In this case, cooperation could take the form of additional withdrawal from the aquifer, but may also include resource-governance programmes to ensure cooperation and sustainable use. In fact, one could envision technical assistance taking the form of support for Syria's nascent inspection and enforcement efforts, including Syria's efforts to reduce groundwater-irrigated areas by introducing new technology (Alabdullah and Nuppenau 2010).

Of course, the mere existence of data will not change management practices. Yet these models can demonstrate expected impacts of such actions, thus leading to a clearer valuation of potential management options.

Note

1. DSI sub-contracted TUMAS to conduct the study.

References

Alabdullah, I. and Nuppenau, E.-A., 2010. Irrigation water policies in Syria: current situation analysis and ccenarios for the future. Paper presented at the World Food System: A Contribution from Europe Conference, Zurich, 14–16 September 2010.

Amos, D., 2010. *Mideast water crisis brings misery, uncertainty*. National Public Radio, January 7.

Anderson, M.P. and Woessner, W.W., 2002. *Applied groundwater modeling: simulation of flow and advective transport*. San Diego: Academic Press.

Assyrian International News Agency, 2000. Syria's water policy targets Assyrian Christians [online]. Available from: http://www.aina.org/releases/syrias.htm [Accessed 14 February 2011].

Bear, J., Beljin, M.S., and Ross, R.R., 1992. *Fundamentals of groundwater modeling*. US EPA Ground water Issue, EPA/540/S-92/005. Ada, Oklahoma: EPA. Available from: http://www.epa.gov/tio/tsp/download/issue13.pdf/

Bredehoeft, J.D., 2002. The water budget myth revisited: why hydrogeologists model. *Ground Water*, 40 (4), 340–345.

Burdon, D.J. and Safadi, C., 1963. Ras-el-Ain: the great karst spring of Mesopotamia: an hydrogeological study. *Journal of Hydrology*, 1 (1), 58–95.

Carpenter, K.E., 1992. Preliminary observations on the effects of the 1991 Gulf War on fisheries. *Marine Pollution Bulletin*, 24 (5), 273–275.

Central Intelligence Agency, 2011. *The world factbook* [online]. Available from: https://www.cia.gov/library/publications/the-world-factbook/ [Accessed 22 February 2011].

Chermark, J.M., Patrick, R.H., and Brookshire, D.S., 2005. Economics of transboundary aquifer management. *Ground Water*, 43 (5), 731–736.

Critchley, W., Siegert, K., and Chapman, C., 1991. A manual for the design and construction of water harvesting schemes for plant production. Rome: Food and Agriculture Organization.

Devlin, J.F. and Sophocleous, M., 2005. The persistence of water budget myth and its relationship to sustainability. *Hydrogeology*, 13 (4), 549–554.

Dimitrov, R.S., 2003. Knowledge, power, and interests in environmental regime formation. *International Studies Quarterly*, 47 (1), 123–150.

Domenico, P.A. and Schartz, F.W., 1998. *Physical and chemical hydrogeology*. 2nd ed. New York: John Wiley & Sons.

FAO, 2010. *FAO's Information system on water and agriculture* [online]. Rome: Food and Agriculture Organization. Available from: http://www.fao.org/nr/water/aquastat/countries/turkey/index.stm [Accessed 11 December 2010].

Fuller, G.E., 2008. *The new Turkish Republic: Turkey as a pivotal state in the Muslim world*. Washington, DC: United States Institute of Peace Press.

Gleick, P.H., 1993. Water and conflict: fresh water resources and international security. *International Security*, 18 (1), 79–112.

Gogu, R.C., *et al.*, 2001. GIS-based hydrogeological databases and groundwater modeling. *Hydrogeology*, 9 (6), 555–569.

Heijde, P., *et al.*, 1985. *Groundwater management: the use of numerical models*. Washington, DC: American Geopyhsical Union.

Hurriyet Daily News, 2009. *He who clears the mines, tills land*, 17 June.

Hurriyet Daily News, 2011. *Syria, Turkey to build dam on border running along Orontes River*, 27 January.

International Campaign to Ban Landmines, 2001. *Landmine monitor report*. New York: Human Rights Watch.

Jarvis, T., 2005. International borders, ground-water flow, and hydroschizophrenia. *Ground Water*, 43 (5), 764–770.

Kattan, Z., 2001. Use of hydrochemistry and environmental isotopes for evaluation of groundwater in the Paleogene Limestone Aquifer of the Ras Al-Ain Area (Syrian Jezireh). *Environmental Geology*, 41 (1–2), 128–144.

Keohane, R.O. and Nye, J.S., 2001. *Power and interdependence*. 3rd ed. New York: Longman.

Kibaroglu, A., *et al.*, 2006. Cooperation on Turkey's transboundary waters. Paper presented at 3rd International Symposium on Transboundary Water Management, Universidad de Castilla-La Mancha, Ciudad Real, Spain, 30 May–2 June 2006.

Kuniansky, E.L., Gomez, F.G., and Gonzalez, S.T., 2004. Effects of aquifer development and changes in irrigation practices on ground-water availability in the Santa Isabel Area, Puerto Rico. *In*: *Water resources investigation report* 03–4303, 65. Reston, VA: US Geological Survey, 1–156.

Li, S.G. and Liu, Q., 2006. Interactive Ground Water (IGW). *Environmental Modelling & Software*, 21 (3), 417–418.

Mende, A., Astorga, A., and Neumann, D., 2007. Strategy for groundwater management in developing countries: a case study in Northern Costa Rica. *Journal of Hydrology*, 334 (1–2), 109–124.

Moore, J.E., 1983. Ground-water models for water resources planning. *GeoJournal*, 7 (5), 453–458.

Oldac, Y., *et al.*, 1971. *Ceylanpinar Plain hydrogeological investigation report*, 60. Ankara: Republic of Turkey General Directorate of State Hydraulic Works.

Olson, R., 1996. The impact of the Southeast Anatolian Project (GAP) on Kurdish nationalism in Turkey. *The International Journal of Kurdish Studies*, 9 (1–2), 95–102.

Puri, S. and Naser, H.E., 2003. Intensive use of groundwater in transboundary aquifers. *In*: M. R. Llamas and E. Custodio, eds, *Intensive use of groundwater: challenges and opportunities*. Leiden: Swets & Zeitlinger, 415–438.

Puri, S., *et al.*, 2001. Internationally shared (transboundary) aquifer resources management: their significance and sustainable management: a framework document. *In*: *IHP-VI, Series in Groundwater* No. 1, 66. Paris: International Hydrological Programme.

Reuveny, R., Mihalache-O'Keef, A.S.., and Quan Li, 2010. The effect of warfare on the environment. *Journal of Peace Research*, 47 (6), 749–761.

Salman, M. and Mualla, W., 2003. The utilization of water resources in Syria: analysis of current situation and future challenges. Paper presented at: Erice International Seminars on Planetary Emergencies, Sicily, 18–26 August, 2003.

Starr, J.R., 1991. Water wars. *Foreign Policy*, 82, 17–36.

Traub, J.,2011. Turkey's rules. *New York Times Magazine*, January 23.

Turan, I., 2003. Water and Turkish foreign policy. *In*: L.G. Martin and D. Keridis, eds, *The Future of Turkish Foreign Policy*. Cambridge, MA: MIT Press, 191–208.

Turkish Engineering, Consulting, and Contracting Co, (TUMAS), 1991. *Ceylanpinar Region groundwater feasibility report*. Ankara: Southeastern Anatolia Project Regional Development Administration.

Turkish Ministry of Foreign Affairs Department of Regional and Transboundary Waters, 1996. Water Issues Between Turkey, Syria and Iraq. Perceptions. *Journal of International Affairs,* 1(2), Available from: http://www.sam.gov.tr/perceptions/Volume1/June-August1996/WATERISSUESBETWEENTURKEYSYRIAANDIRAQ.pdf [Accessed 22 June 2011].

Watkins, D.W., 1996. Use of Geographic Information Systems in ground-water flow modeling. *Journal of Water Resources Planning and Management*, 122 (2), 88–96.

Watson, I. and Comert, Y., 2009. *Turkey, Iraq and Syria in water crisis summit*. CNN, 3 September.

Index

Note: Figures are indicated by *italics*. Tables are indicated by **bold**. Endnotes are indicated by the page number followed by 'n' and the endnote number e.g., 20n1 refers to endnote 1 on page 20.